HOME LINKS

Everyday
Mathematics®

The University of Chicago School Mathematics Project

Mc
Graw
Hill
Education

The University of Chicago School Mathematics Project

Max Bell, Director, *Everyday Mathematics* First Edition; James McBride, Director, *Everyday Mathematics* Second Edition; Andy Isaacs, Director, *Everyday Mathematics* Third, CCSS, and Fourth Editions; Amy Dillard, Associate Director, *Everyday Mathematics* Third Edition; Rachel Malpass McCall, Associate Director, *Everyday Mathematics* CCSS and Fourth Editions; Mary Ellen Dairyko, Associate Director, *Everyday Mathematics* Fourth Edition

Authors
Max Bell, John Bretzlauf, Amy Dillard, Robert Hartfield, Andy Isaacs, Rebecca Williams Maxcy, James McBride, Kathleen Pitvorec, Peter Saecker, Robert Balfanz*, William Carroll*, Sheila Sconiers*

*First Edition Only

Fourth Edition Grade 4 Team Leader
Rebecca Williams Maxcy

Writers
Meg Schleppenbach Bates, Randee Blair, Kristin Fitzgerald, Carla LaRochelle, Sara A. Snodgrass

Open Response Team
Catherine R. Kelso, Leader; Judith S. Zawojewski, Andy Carter, John Benson

Differentiation Team
Ava Belisle-Chatterjee, Leader; Jean Capper, Martin Gartzman, Barbara Molina

Digital Development Team
Carla Agard-Strickland, Leader; John Benson, Gregory Berns-Leone, Juan Camilo Acevedo

Virtual Learning Community
Meg Schleppenbach Bates, Cheryl G. Moran, Margaret Sharkey

Technical Art
Diana Barrie, Senior Artist; Cherry Inthalangsy

UCSMP Editorial
Don Reneau, Senior Editor; Elizabeth Olin, Kristen Pasmore, Lucas Whalen

Field Test Coordination
Denise A. Porter

Field Test Teachers
Kindra Arwood, Tiffany N. Harper, Brian A. Herman, Tonya Howell, Amy Jacobs, Amy Jarrett-Clancy, Kari Lehman, Stephanie Rogers, Jenna Rose Ryan, Rachel Schrader, JoAnn Tennenbaum, Robin Zogby

Digital Field Test Teachers
Colleen Girard, Michelle Kutanovski, Gina Cipriani, Retonyar Ringold, Catherine Rollings, Julia Schacht, Christine Molina-Rebecca, Monica Diaz de Leon, Tiffany Barnes, Andrea Bonanno-Lersch, Debra Fields, Kellie Johnson, Elyse D'Andrea, Katie Fielden, Jamie Henry, Jill Parisi, Lauren Wolkhamer, Kenecia Moore, Julie Spaite, Sue White, Damaris Miles, Kelly Fitzgerald

Contributors
William Baker, John Benson, Jeanne Di Domenico, Jim Flanders, Lila Schwartz, Penny Williams

Center for Elementary Mathematics and Science Education Administration
Martin Gartzman, Executive Director; Meri B. Fohran, Jose J. Fragoso, Jr., Regina Littleton, Laurie K. Thrasher

External Reviewers
The *Everyday Mathematics* authors gratefully acknowledge the work of the many scholars and teachers who reviewed plans for this edition. All decisions regarding the content and pedagogy of *Everyday Mathematics* were made by the authors and do not necessarily reflect the views of those listed below.

Elizabeth Babcock, California Academy of Sciences; Arthur J. Baroody, University of Illinois at Urbana-Champaign and University of Denver; Dawn Berk, University of Delaware; Diane J. Briars, Pittsburgh, Pennsylvania; Kathryn B. Chval, University of Missouri-Columbia; Kathleen Cramer, University of Minnesota; Ethan Danahy, Tufts University; Tom de Boor, Grunwald Associates; Louis V. DiBello, University of Illinois at Chicago; Corey Drake, Michigan State University; David Foster, Silicon Valley Mathematics Initiative; Funda Gönülateş, Michigan State University; M. Kathleen Heid, Pennsylvania State University; Natalie Jakucyn, Glenbrook South High School, Glenview, IL; Richard G. Kron, University of Chicago; Richard Lehrer, Vanderbilt University; Susan C. Levine, University of Chicago; Lorraine M. Males, University of Nebraska-Lincoln; Dr. George Mehler, Temple University and Central Bucks School District, Pennsylvania; Kenny Huy Nguyen, North Carolina State University; Mark Oreglia, University of Chicago; Sandra Overcash, Virginia Beach City Public Schools, Virginia; Raedy M. Ping, University of Chicago; Kevin L. Polk, Aveniros LLC; Sarah R. Powell, University of Texas at Austin; Janine T. Remillard, University of Pennsylvania; John P. Smith III, Michigan State University; Mary Kay Stein, University of Pittsburgh; Dale Truding, Arlington Heights District 25, Arlington Heights, Illinois; Judith S. Zawojewski, Illinois Institute of Technology

Note
Many people have contributed to the creation of *Everyday Mathematics*. Visit http://everydaymath.uchicago.edu/authors/ for biographical sketches of *Everyday Mathematics* 4 staff and copyright pages from earlier editions.

www.everydaymath.com

Send all inquiries to:
McGraw-Hill Education
8787 Orion Place
Columbus, OH 43240

ISBN: 978-0-02-137966-8
MHID: 0-02-137966-1

Printed in the United States of America.

7 8 9 10 LHS 20 19

Contents

Place Value in Whole Numbers

Family Note In this lesson your child explored the relationships between place values in numbers. Have your child read each number below. Examine the digit 6 in each number.

Hundred-Thousands	Ten-Thousands	Thousands	Hundreds	Tens	Ones
600,000	60,000	6,000	600	60	6

When the digit 6 moves left one place, its value becomes 10 times as large as it was in the previous place. For example, 60 is 10 times as large as 6, and 600 is 10 times as large as 60.

①   **a.** The 8 in 203,810 is worth _____.   **b.** The 6 in 56,143 is worth _____.

SRB
78-79

   c. The 7 in 573,090 is worth _____.   **d.** The 1 in 140,007 is worth _____.

② How does the value of the digit 4 in 489 differ from the value of the digit 4 in 5,741?

_____

_____

③   **a.** The value of 8 in 56,982 is _____ times as large as the value of 8 in 156,408.

   b. The value of 8 in 800 is _____ times as large as the value of 8 in 80.

   c. The value of 9 in 4,934 is _____ times as large as the value of 9 in 1,290.

④   **a.** Write the number that has . . .
     7 in the thousands place
     6 in the ten-thousands place
     5 in the hundreds place
     8 in the ones place
     3 in the tens place

     _____ _____, _____ _____ _____

   b. On the back of this page, write this number in words.

Practice

⑤ $9 + 8 =$ _____   ⑥ $7 + 8 =$ _____   ⑦ $30 + 80 =$ _____

⑧ _____ $= 50 + 40$ ⑨ _____ $= 17 + 94$ ⑩ $158 + 93 =$ _____

Introduction to *Fourth Grade Everyday Mathematics*

Welcome to *Fourth Grade Everyday Mathematics*, part of an elementary school mathematics curriculum developed by the University of Chicago School Mathematics Project (UCSMP).

Fourth Grade Everyday Mathematics emphasizes the following content:

Operations and Algebraic Thinking Investigating methods for solving problems involving mathematics in everyday situations; solving multistep problems involving the four operations; using estimation to check the reasonableness of answers; exploring properties of numbers such as multiples, factor pairs, prime and composite; and designing, exploring, and using geometric and number patterns.

Numbers and Operations in Base Ten Reading, writing, comparing, and ordering whole numbers; exploring addition, subtraction, multiplication, and division methods; inventing individual procedures and methods.

Number and Operations—Fractions Developing an understanding of fraction equivalence; exploring addition and subtraction of fractions with like denominators and multiplication of fractions by whole numbers; and reading, writing, comparing, and ordering fractions and decimals.

Measurement and Data Exploring metric and U.S. customary measurement systems and converting from larger units to smaller units within a single system; applying formulas to find the perimeters and areas of rectangles; developing an understanding of angles and angle measurement; and representing and interpreting data on line plots.

Geometry Drawing and identifying geometric properties and identifying these properties in polygons; recognizing and drawing a line of symmetry; identifying symmetric figures.

Everyday Mathematics provides you with many opportunities to monitor your child's progress and participate in your child's experience of mathematics. Throughout the year you will receive Family Letters to keep you informed of the mathematical content your child will be studying in each unit. Each letter includes a vocabulary list, suggested Do-Anytime Activities for you and your child, and an answer guide to selected Home Link (homework) activities. You will enjoy seeing your child's confidence and comprehension soar as he or she connects mathematics to everyday life.

3

This unit reviews and extends mathematical content developed in *Third Grade Everyday Mathematics.* In Unit 1, students will explore the following concepts:

Place Value in Whole Numbers Students review place-value concepts and explore numbers in the ten-thousands and hundred-thousands. They will read, write, compare, and order these numbers. Students will also use population data from U.S. cities to practice rounding and comparison techniques.

Computation Students practice mental and paper-and-pencil methods of computation, as well as using a calculator. They will decide which tool is most appropriate for solving a particular problem.

Students explore a new strategy for adding and subtracting multidigit whole numbers and compare different methods. They will realize that often the same result may be obtained in multiple ways.

Students use estimation to assess the reasonableness of answers as they work with multistep number stories using a letter for the unknown.

Measurement and Data Students review the concept of perimeter and then develop and apply formulas for finding the perimeters of rectangles.

Students convert between customary units of length (yards, feet, inches) and solve number stories involving conversions.

Geometry Students examine definitions and properties of 2-dimensional shapes and the relationships among them.

Please keep this Family Letter for reference as your child works through Unit 1.

Vocabulary

Important terms in Unit 1:

acute angle An angle with a measure greater than 0° and less than 90°.

acute angle

angle A figure that is formed by two rays or two line segments with a common endpoint.

base 10 Our number system in which each place in a number has a value 10 times as large as the place to its right and $\frac{1}{10}$ the place to its left.

digit One of the number symbols 0, 1, 2, 3, 4, 5, 6, 7, 8, and 9 in the standard base-10 system.

endpoint A point at the end of a line segment or ray.

line segment *LT*

expanded form A way of writing a number as the sum of the values of each digit. For example, the expanded form of 356 is 300 + 50 + 6.

intersect To share a common point or points.

intersecting
lines and segments

intersecting
planes

line A straight path that extends infinitely in opposite directions.

line *PR*

line segment A straight path joining two points, which are called endpoints.

line segment *AB*

obtuse angle An angle that has a measure greater than 90° and less than 180°.

obtuse angle

parallel Lines, line segments, or rays in the same plane are parallel if they never cross or meet, no matter how far they are extended in either direction.

parallel lines

parallelogram A quadrilateral that has two pairs of parallel sides. Opposite sides of a parallelogram have equal lengths, and its opposite angles have the same measure.

perimeter The distance around the boundary of a 2-dimensional figure.

perimeter = 1 cm + 3 cm + 1 cm + 3 cm = 8 cm

perpendicular Crossing or meeting at right angles. Lines, rays, or line segments that cross or meet at right angles are perpendicular.

perpendicular
lines

place value The value given to a digit according to its position, or place, in a number. The chart on the next page shows the value of each digit in 24,815.

point An exact location in space. Lines have an infinite number of points on them.

ray A straight path that extends infinitely from an endpoint. A ray is named using the letter label of its endpoint followed by the letter label of another point on the ray.

ray *MN*

right angle An angle that measures exactly 90°.

right triangle A triangle that contains a right angle.

right triangle

U.S. traditional addition A paper-and-pencil method for adding multidigit numbers in which the addends are stacked vertically with place values aligned and the digits in each column are added, working column by column from the right. The tens digit, if any, from a column sum is "carried" to the top of the next column to the left and is added with the digits in that column.

U.S. traditional subtraction A paper-and-pencil method for subtracting multidigit numbers. The minuend (number from which another is subtracted) and subtrahend (number being subtracted) are stacked vertically with place values aligned and the digits in each column are subtracted, working column by column from the

right. If a digit in the subtrahend is larger than the corresponding digit in the minuend, a 10 is "borrowed" from the next column to the left.

vertex A point at which the rays of an angle or the sides of a geometric figure meet.

whole numbers The numbers 0, 1, 2, 3, 4, and so on.

Ten-Thousands	Thousands	Hundreds	Tens	Ones
2	4	8	1	5
The value of the 2 is 20,000.	The value of the 4 is 4,000.	The value of the 8 is 800.	The value of the 1 is 10.	The value of the 5 is 5.

Place-value chart

Do-Anytime Activities

To work with your child on concepts taught in this unit, try these activities:

1. Have your child locate big numbers in newspapers and other sources and ask him or her to read them to you. Or read the numbers and have your child write them down.

2. Help your child look up the populations and land areas of the state and city in which you live and compare them with the populations and areas of other states and cities.

3. Together, write five multidigit numbers in order from smallest to largest.

4. Model real-life uses of estimation for your child. For example, when you are shopping, round the cost of several items up to the nearest dollar and add to estimate their total cost.

5. Help your child discover everyday uses of geometry found in art, architecture, jewelry, toys, and so on.

Building Skills through Games

Throughout the school year, students will play mathematics games as a way to practice a variety of arithmetic skills. Playing games turns practice into a fun thinking activity. Games in this unit provide practice with place value, addition, and subtraction. They require very few materials, so you and your child can play them at home.

Addition Top-It (**Advanced Version**) See *Student Reference Book*, page 275. This variation of *Addition Top-It* provides practice adding numbers through the thousands.

Fishing for Digits See *Student Reference Book*, page 259. This game provides practice identifying digits and their values, as well as adding and subtracting.

Number Top-It See *Student Reference Book*, page 269. This game provides practice working with place value through the hundred-thousands.

Subtraction Top-It (**Advanced Version**) See *Student Reference Book*, page 275. This variation of *Subtraction Top-It* provides practice subtracting numbers through the thousands.

As You Help Your Child with Homework

As your child brings assignments home, you may want to go over the instructions together, clarifying them as necessary. The answers listed below will guide you through the Home Links for this unit.

Home Link 1-1

1. **a.** 800   **b.** 6,000   **c.** 70,000   **d.** 100,000
3. 10; 10; 10
4. 67,538 ; Sixty-seven thousand, five hundred thirty-eight

Home Link 1-2

3a. Uganda. Both have the same number of ten-thousands, but Uganda's area has 3 thousands and Laos's area has 1 thousand.

3b. 93,100 > 91,400

Home Link 1-3

1. Chicago Sky: 18,000
   Connecticut Sun: 10,000
   Indiana Fever: 18,000
   Los Angeles Sparks: 13,000
   Minnesota Lynx: 19,000
   Phoenix Mercury: 18,000
   Seattle Storm: 17,000
   Tulsa Shock: 18,000
   Washington Mystics: 20,000

3. Wyoming: 600,000
   Vermont: 600,000
   North Dakota: 700,000
   Alaska: 700,000
   South Dakota: 800,000

Home Link 1-4

1. Baseball

3. 3,000,000; 4,000,000; 4,000,000; 3,000,000; 2,000,000

5. 2,370,794 < 3,565,718

Home Link 1-5

1. **a.** No. Sample answer: I rounded the times to the tens place and added: $20 + 40 = 60$ and $60 + 20 = 80$.

   b. Sample answer: The numbers were all close to a multiple of 10. I just needed to know if they added up to more or less than 60.

2. **a.** No. Sample answer: I used close-but-easier numbers. I rounded 31 to 30 and 24 to 25. $30 + 25 = 55$; $100 - 55 = 45$.

   b. Sample answer: The numbers were all close to friendly numbers, so I decided to go with close-but-easier numbers.

Home Link 1-6

Estimates vary.

1. 150 pounds; 144 pounds; Sample answer: $700 - (176 + 250 + 130) = 144$; Yes. Sample answer: My estimate was 150 pounds, which is close to my answer.

2. 300 pounds; 272 pounds; Sample answer: $(491 - (175 + 180)) * 2 = 272$; Yes; Sample answer: My estimate was 300 pounds, which is close to 272 pounds.

Home Link 1-7

Estimates vary.

1. 82; $40 + 50 = 90$ 3. 1,673; $800 + 900 = 1,700$

5. 2,074; $300 + 1,800 = 2,100$ 7. 2,800; 3,000

Home Link 1-8

1. 100 balls

2. 730 balls; Sample answer: $7 * 100$ balls in a carton $+ 3 * 10$ balls in a box $= 730$ balls

3. Sample answer: The number of cartons is like the digit in the 100s place, and the number of boxes is like the digit in the 10s place.

Home Link 1-9

Estimates vary.

1. 47; $90 - 40 = 50$ 3. 319; $500 - 200 = 300$

5. 795; $2,000 - 1,000 = 1,000$ 7. 1,034

Home Link 1-10

1. 12; 72; 96; 144   3. 27 feet

5. Four thousand, eight hundred fifty-seven

Home Link 1-11

3. **a.**

   b. No. A ray's endpoint must be listed first when naming a ray.

4. $\overrightarrow{WX}$ (or $\overrightarrow{XW}$) is parallel to $\overleftrightarrow{ST}$ (or $\overleftrightarrow{TS}$).

Home Link 1-12

1.

3. Both have right angles and perpendicular sides. They have a different number of sides and right angles.

4. **a.**   **b.** *E*   **c.** $\angle FED$

5.

Home Link 1-13

1. 30 feet   2. 42 inches   3. 116 feet

4. 108 inches   5. 6 feet

7. 900,000; 900,000

Country Sizes

This table shows the sizes of
10 countries measured in square miles.

SRB
81

Use a place-value tool to help you answer
the questions.

Country	Area (in square miles)
Algeria	919,600
Colombia	439,700
Ethiopia	426,400
Egypt	386,700
Greece	50,900
Iran	636,400
Laos	91,400
Peru	494,200
Uganda	93,100

① Read the numbers to someone at home.

② Which is the largest country listed?

_____

The smallest? _____

Source: worldatlas.com (All data rounded to nearest hundred.)

③ Compare the areas of Laos and Uganda.

a. Which country has the larger area? _____ How do you know?

_____

_____

b. Write a comparison number sentence. _____

④ Order the countries from largest
area to smallest area.

Country	Area (in square miles)

Practice

⑤ $140 - 60 =$ _____ ⑥ _____ $= 57 - 39$ ⑦ $115 - 86 =$ _____

Rounding

(1) Round the seating capacities in the table below to the nearest thousand.

Women's National Basketball Association (WNBA) Seating Capacity of Home Courts		
Team	**Seating Capacity**	**Rounded to the Nearest 1,000**
Chicago Sky	17,500	
Connecticut Sun	9,518	
Indiana Fever	18,165	
Los Angeles Sparks	13,141	
Minnesota Lynx	19,356	
Phoenix Mercury	18,422	
Seattle Storm	17,072	
Tulsa Shock	17,839	
Washington Mystics	20,308	

Source: www.wnba.com

(2) Look at your rounded numbers. Which teams' arenas have about the same capacity?

_____

(3) Round the population figures in the table below to the nearest hundred-thousand.

U.S. States with the Five Smallest Populations (2010 Census)		
State	**Population**	**Rounded to the Nearest 100,000**
Wyoming	563,626	
Vermont	626,011	
North Dakota	699,628	
Alaska	731,449	
South Dakota	833,354	

Practice

(4) _____ = 60 + 60   (5) _____ = 54 + 59   (6) 185 + 366 = _____

11

Professional Sports Attendance

The table below shows the attendance for various 2013–2014 professional sports teams. Use the table and a place-value tool to answer the questions.

SRB
81,
85-87

	Chicago*	New York*†	Philadelphia	Boston	Washington
Hockey	927,545	738,246	813,411	720,165	740,240
Baseball	2,882,756	3,542,406	3,565,718	3,043,003	2,370,794

Source: ESPN NHL Attendance report 2013–2014 and ESPN MLB Attendance report 2012
*Baseball attendance is for the Chicago Cubs and the New York Yankees.
†Hockey attendance is for the New York Rangers.

(1) Which sport had the greater attendance? _____

(2) Round the attendance at the hockey games.

	Nearest 100,000	Nearest 10,000
Chicago		
New York		
Philadelphia		
Boston		
Washington		

(3) Round the attendance for each baseball team to the nearest million.

Chicago: _____   New York: _____

Philadelphia: _____   Boston: _____

Washington: _____

(4) List the cities in order from greatest to least hockey attendance.

_____

(5) Write a number sentence comparing the greatest and least baseball attendances. Use <, >, or =.

_____

Practice

(6) 210 − 150 = _____   (7) 140 − 80 = _____   (8) 93 − 58 = _____

13

Using Estimation Strategies

Family Note Today students explored different ways of estimating: **rounding** (in which all numbers are rounded to a particular place value), **front-end estimation** (all digits to the right of the greatest place value become zeros), and using **close-but-easier numbers** (numbers are rounded to a number that is close in value and easy to work with). While all methods of estimation are equally valid, some may be more helpful than others for answering specific kinds of questions.

Read the number stories. Choose an appropriate estimation strategy.

SRB
82-89

① On the walk home from school, Meg stopped at the library for 22 minutes and at her grandmother's house for 38 minutes. She spent 17 minutes walking. She left at 3:00 and was supposed to be home by 4:00.

    a. Did Meg make it home on time? _____ How did you get your answer?

    _____

    _____

    b. Why did you choose your estimation strategy? _____

    _____

    _____

② You and two friends need to make 100 tacos for a party. You have made 31 tacos. Your friend Chris has made 24 tacos. Your friend Pat thinks he needs to make at least 60 tacos to have enough for the party.

    a. Is Pat correct? _____ How did you get your answer?

    _____

    _____

    b. Why did you choose your estimation strategy? _____

    _____

    _____

Practice

③ $31 + 51 =$ _____   ④ $45 + 64 =$ _____   ⑤ $252 + 144 =$ _____

15

Animal Number Stories

SRB
82-84

Estimate. Then solve each number story.

(1) The zoo needs to move four animals in a truck that can carry only 700 pounds. A leopard can weigh up to 176 pounds. A warthog can weigh up to 250 pounds. A chimpanzee can weigh as much as 130 pounds. What is the maximum weight that the fourth animal can be?

Estimate: About _____ pounds

Answer: _____ pounds

Number model with answer: _____

Does your answer make sense? _____ How do you know?

_____

_____

(2) The combined weight of a mountain lion, an orangutan, and a wolf can be as much as 491 pounds. If the wolf weighs 175 pounds and the orangutan weighs 180 pounds, how much do *two* mountain lions weigh?

Estimate: About _____ pounds

Answer: _____ pounds

Number model with answer: _____

Does your answer make sense? _____ How do you know?

_____

_____

Source: maximum animal weights from www.nationalgeographic.com

Practice

(3) $5 + 8 =$ _____

(4) $9 + 6 =$ _____

(5) $70 + 50 =$ _____

(6) _____ $= 80 + 50$

(7) $67 + 94 =$ _____

(8) _____ $= 425 + 275$

17

U.S. Traditional Addition

Family Note In today's lesson students were introduced to U.S. traditional addition. The steps are listed below.

SRB
92-93

Step 1

Add the 1s: 9 + 7 = 16.

16 ones = 1 ten and 6 ones

Write 6 in the 1s place below the line.

Write 1 above the digits in the 10s place.

$$\begin{array}{r} \overset{1}{7}\,9 \\ +\ 4\ 7 \\ \hline 6 \end{array}$$

Step 2

Add the 10s: 7 + 4 + 1 = 12.

12 tens = 1 hundred + 2 tens

Write 2 in the 10s place below the line.

Write 1 in the 100s place below the line.

$$\begin{array}{r} \overset{1}{7}\,9 \\ +\ 4\ 7 \\ \hline 1\ 2\ 6 \end{array}$$

Make an estimate. Write a number model to show what you did. Then solve using U.S. traditional addition. Compare your answer with your estimate to see if your answer makes sense.

①
$$\begin{array}{r} 3\ 6 \\ +\ 4\ 6 \end{array}$$

Estimate: _____

②
$$\begin{array}{r} 4\ 7 \\ +\ 9\ 5 \end{array}$$

Estimate: _____

③ 784 + 889 =

Estimate: _____

④
$$\begin{array}{r} 6\ 8\ 9 \\ +\ 8\ 3\ 9 \end{array}$$

Estimate: _____

⑤ 279 + 1,795 =

Estimate: _____

⑥ 3,746 + 6,255 =

Estimate: _____

Practice

⑦ Round 2,787 to the nearest . . .

hundred _____   thousand _____

⑧ Round 54,681 to the nearest . . .

thousand _____   ten-thousand _____

Grouping by Multiples of 10

Alfie is ordering table tennis balls for the recreation center. A box holds 10 balls. A carton of table tennis balls holds 10 boxes.

Box of table tennis balls

Carton of table tennis balls

(1) How many table tennis balls are in one carton? _____

(2) Alfie ordered 7 cartons and 3 boxes of table tennis balls. How many balls did he order? _____

Show how you know your answer is correct.

(3) Explain how the cartons and boxes for table tennis balls are like the digits for numbers in our base-10 number system.

Practice

(4) 440 + 294 = _____

(5) 166 + 707 = _____

(6) _____ = 425 + 886

(7) 1,474 + 529 = _____

21

U.S. Traditional Subtraction

Family Note In today's lesson students were introduced to U.S. traditional subtraction. The process is shown below for the problem 653 − 387.

SRB
82-84,
100-101

Step 1:

Start with the ones. Trade 1 ten for 10 ones. Subtract the ones.

100s	10s	1s
	4	13
6	5̸	3̸
− 3	8	7
		6

Step 2:

Go to the tens. Trade 1 hundred for 10 tens. Subtract the tens.

100s	10s	1s
	14	
5	4̸	13
6̸	5̸	3̸
− 3	8	7
	6	6

Step 3:

Go to the hundreds. We don't need to regroup, so just subtract.

100s	10s	1s
	14	
5	4̸	13
6̸	5̸	3̸
− 3	8	7
2	6	6

Make an estimate. Write a number model to show what you did. Then solve using U.S. traditional subtraction. Compare your answer with your estimate to see whether your answer makes sense.

①
```
    8 5
  − 3 8
```
Estimate: _____

②
```
    6 1 3
  − 2 4 9
```
Estimate: _____

③ 506 − 187 = _____

Estimate: _____

④ 951 − 695 = _____

Estimate: _____

⑤
```
   1, 5 4 4
   −   7 4 9
```
Estimate: _____

⑥ 7,003 − 4,885 =

_____

Estimate: _____

Practice

⑦ 740 + 294 = _____

⑧ 2,566 + 807 = _____

Snake Lengths

Use the measurement scales to solve the problems.

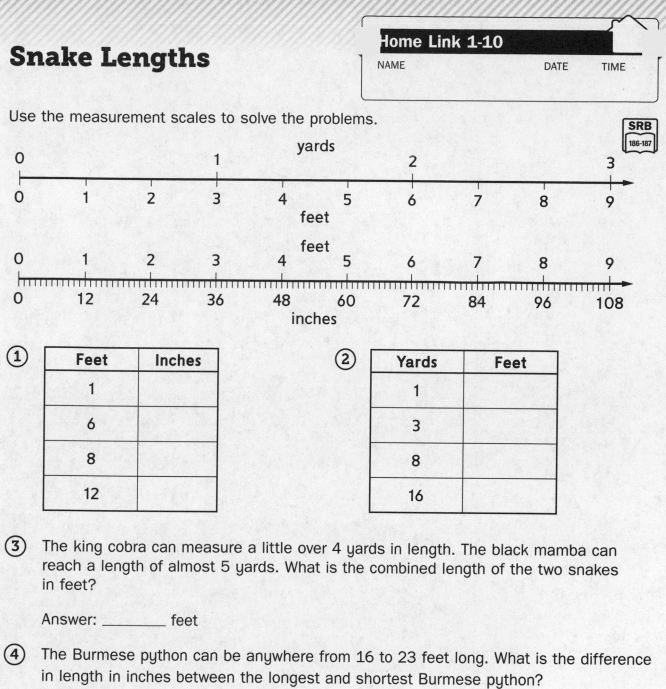

①
Feet	Inches
1	
6	
8	
12	

②
Yards	Feet
1	
3	
8	
16	

③ The king cobra can measure a little over 4 yards in length. The black mamba can reach a length of almost 5 yards. What is the combined length of the two snakes in feet?

Answer: _____ feet

④ The Burmese python can be anywhere from 16 to 23 feet long. What is the difference in length in inches between the longest and shortest Burmese python?

Answer: _____ inches

Practice

⑤ Write 4,857 in words.

_____

⑥ Write 14,066 in words.

_____

25

Line Segments, Lines, and Rays

① List at least 5 things in your home that remind you of line segments.

SRB
226-227,
230-231

_____

_____

_____

Use a straightedge to complete Problems 2 and 3.

② **a.** Draw and label line *EF*.   **b.** Draw and label line segment *EF*.

c. Explain how your drawings of line *EF* and line segment *EF* are different.

_____

_____

③ **a.** Draw and label ray *SR*.

b. Anita says ray *SR* can also be called ray *RS*. Do you agree? Explain.

_____

_____

④

W   X

S   T

Name the parallel line segments.

_____

Practice

⑤
```
  9 6 4
- 3 4 8
```

⑥
```
  6 6 2
- 4 9 7
```

⑦
```
  2, 4 2 3
- 1, 4 9 1
```

27

Angles and
Quadrilaterals

Use a straightedge to draw the geometric figures.

SRB
228-229,
233

(1) Draw 2 examples of a rectangle.

(2) Draw 2 examples of a right triangle.

(3) How are the shapes in Problems 1 and 2 similar? How are they different?

_____

_____

_____

(4) **a.** Draw right angle *DEF*.

(5) Draw an angle that is larger than a right angle. Label the vertex *K*.

b. What is the vertex of the angle? Point _____

c. What is another name for ∠*DEF*? _____

Practice

Use U.S. traditional subtraction.

(6) _____ = 756 − 348

(7) 700 − 450 = _____

(8) 7,942 − 3,887 = _____

29

Finding the Perimeter

Family Note In class, students developed some rules, or *formulas,* for finding the perimeter of a rectangle. Here are three possible formulas:

- Add the measures of the four sides: perimeter of a rectangle = length + length + width + width. This formula can be abbreviated as: $p = l + l + w + w$.

- Add the two given sides and double the sum: perimeter of a rectangle = 2 * (length + width). This formula can be abbreviated as: $p = 2 * (l + w)$.

- Double the length, double the width, and then add: perimeter of a rectangle = (2 * length) + (2 * width). This formula can be abbreviated as: $p = 2l + 2w$.

In all of the formulas, the letter *p* stands for the *perimeter of a rectangle,* the letter *l* stands for the *length of the rectangle,* and the letter *w* stands for the *width of the rectangle.*

Find the perimeters of the rectangles below.

SRB
200

(1)

7 feet

8 feet

Perimeter: _____ feet

(2)

9 inches

12 inches

Perimeter: _____ inches

(3)

22 feet

36 feet

Perimeter: _____ feet

(4)

2 feet

30 inches

Perimeter: _____ inches

(5) The perimeter of a garden is 42 feet. The length is 15 feet. What is the width?

Width: _____ feet

Practice

Round each number to the nearest ten-thousand and hundred-thousand.

(6) 421,492 _____ _____

(7) 895,531 _____ _____

31

Multiplication and Multiplicative Comparison

In Unit 2 students build on prior work multiplying whole numbers. The focus is on multiplication in a variety of contexts, including rectangular-array patterns and work with factors, factor pairs, multiples, prime numbers, and composite numbers.

This unit introduces the concept of multiplicative comparison, or using multiplication to compare one quantity to another. Take the following number story: *Mike earned $4. Sue earned 7 times as much as Mike*. Here Sue's earnings are compared to Mike's as being 7 times as much. Based on this comparison, we can find how much Sue earned ($4 * 7 = $28).

Measurement work in Unit 2 is tied to multiplication. Working with units of time, students multiply to convert from hours to minutes and minutes to seconds. They are introduced to the area formula for rectangles, $A = l * w$, in which A is area, l is length, and w is width.

Applying the formula for the area of a rectangle:

$A = 15$ cm² $w = 3$ cm

$l = 5$ cm

Students also work with patterns found in square numbers, multiples, factors, and "What's My Rule?" tables. They practice looking more deeply into patterns by identifying ones that are apparent but are not stated in the rule. For example, students may notice in the pattern based on the rule *multiply a number by itself* that every other square number is even.

Classifying Geometric Figures; Symmetry

Students build on their study of geometry in Unit 1 by identifying properties of shapes. They explore the properties of angles and triangles by identifying right, obtuse, and acute angles in triangles. Students begin work with classification, an important geometry skill, by sorting quadrilaterals according to the number of pairs of parallel sides.

acute angle

right angle   acute angle

Identifying properties of right triangles

The trapezoid has one pair of parallel sides: $\overline{AB}$ and $\overline{CD}$.

Symmetry is another focus in Unit 2. Symmetry is found in natural objects like flowers, insects, and the human body, as well as in buildings, furniture, clothing, and paintings.

Please keep this Family Letter for reference as your child works through Unit 2.

Vocabulary

Important terms in Unit 2:

array An arrangement of objects in a regular pattern, usually in rows and columns.

composite number A counting number that has more than two different *factors*. For example, 4 is a composite number because it has three *factors*: 1, 2, and 4.

factor One of two or more numbers that are multiplied to give a *product*. For example, $4 * 5 = 20$; so 20 is the *product*, and 4 and 5 are the factors.

formula A general rule for finding the value of something. A formula is often written using letters to stand for the quantities involved. For example, the formula for the area of a rectangle may be written as $A = l * w$, where A represents the area of the rectangle, l represents its length, and w represents its width.

line of symmetry A line drawn through a figure that divides the figure into two parts that are mirror images of each other. The two parts look alike but face in opposite directions.

line of symmetry

line symmetry A figure has line symmetry if a line can be drawn dividing it into two parts that are mirror images of each other. The two parts look alike but face in opposite directions.

multiple A *product* of a number and a counting number. The multiples of 7, for example, are 7, 14, 21, 28, and so on.

prime number A counting number greater than 1 that has exactly two *factors*: itself and 1. For example, 5 is a prime number because its only *factors* are 5 and 1.

product The result of multiplying two numbers called *factors*. For example, in $4 * 3 = 12$, the product is 12.

rectangular array An arrangement of objects into rows and columns that form a rectangle. All rows and columns must be filled. Each row has the same number of objects and each column has the same number of objects.

column
row

square array An arrangement of objects into rows and columns that form a square. All rows and columns must be filled. All of the rows and all of the columns have the same number of objects, making the number of rows and columns equal. A square array can be a representation of a *square number*.

square number A number that is the product of a counting number multiplied by itself. For example, 25 is a square number because $5 * 5 = 25$. The square numbers are 1, 4, 9, 16, 25, and so on.

Do-Anytime Activities

To work with your child on concepts taught in this unit, try these activities:

1. Ask your child to list the first 5 or 10 multiples of different 1-digit numbers.

2. Help your child recognize real-world examples of right angles, such as the corner of a book, and of parallel lines, such as railroad tracks.

3. Encourage your child to look for symmetrical objects and if possible to collect pictures of symmetrical objects from magazines and newspapers. For example, the right half of the printed letter T is the mirror image of the left half.

Building Skills through Games

In this unit your child will play the following games to develop understanding of factors and multiples.

Buzz* and *Bizz-Buzz See *Student Reference Book,* page 252. *Buzz* provides practice finding multiples of whole numbers. *Bizz-Buzz* provides practice finding common multiples of two whole numbers.

As You Help Your Child with Homework

As your child brings assignments home, you may want to go over the instructions together, clarifying them as necessary. The answers listed below will guide you through the Home Links for this unit.

Home Link 2-1

1. 2 * 2; 16; 5 * 5; 6 * 6

3. Sample answers: The product of two even factors is even; the product of two odd factors is odd.

5 **a.** 5 * 5 = 25 **b.** 5 * 5 = 25 shows the same number of rows and columns.

7. 11; 37; 63; + 26

Home Link 2-2

1.

9 * 4 = 36; 36

3. 8 * 6 = 48; 48   5. 9 * 6 = 54; 54

7. 47

Home Link 2-3

1. **9:** 1 * 9 = 9, 9 * 1 = 9, 3 * 3 = 9; 1 and 9, 3 and 3; **10:** 1 * 10 = 10, 10 * 1 = 10, 2 * 5 = 10, 5 * 2 = 10; 1 and 10, 2 and 5; **17:** 1 * 17 = 17, 17 * 1 = 17; 1 and 17; **40:** 1 * 40 = 40, 2 * 20 = 40, 4 * 10 = 40, 5 * 8 = 40, 8 * 5 = 40, 10 * 4 = 40, 20 * 2 = 40, 40 * 1 = 40; 1 and 40, 2 and 20, 4 and 10, 5 and 8

3. 2,863   5. 2,182

Home Link 2-4

1. 4, 8, 12, 16, 20

3. **a.** 3, 6, 9, 12, 15, 18, 21, 24, 27, 30

   b. 5, 10, 15, 20, 25, 30, 35, 40, 45, 50

   c. 15 and 30

5. No. 35 cannot be divided evenly by 6.

7. 36; 60; 84; + 12   9. 69; 35; 1; − 17

Home Link 2-5

1. 1, (11); prime

3. 1, (2,) (3,) 4, 6, 8, 12, 24; composite

5. 1, (2,)(3,) 4, 6, 9, 12, 18, 36; composite

7. 1, (2,)(5,) 10, 25, 50; composite

Home Link 2-6

1. 9 grapefruits
3. Answers vary.
5. thirty thousand, forty-one
7. nine million, ninety thousand, five hundred six

Home Link 2-7

1. 240, 480, 660
3. 1,020
4. 47
5. 14,220
7. 7,424
9. 7,298

Home Link 2-8

1. $n = 7 * 9$; 63
3. $32 = 4 * x$; 8
5. 399
7. 2,149

Home Link 2-9

1. Answers vary; Sample answer: $6 * 9 = m$; 54
5. $50,000 + 6,000 + 30 + 7$
7. $700,000 + 10,000 + 6,000 + 300 + 5$

Home Link 2-10

1. C, D
3. C, D
5. C, D, E, F
7. A, B, E, F
9. 1, 2, 3, 4, 6, 12

Home Link 2-11

1. Sample answer:

3. Sample answer:

5. 150
7. 480

Home Link 2-12

1.

3. Answers vary.
5. 7,171
7. 2,595

Home Link 2-13

1. 3; 5; 36; 54

Sample answer: If you add the digits of each of the multiples of 9, the sum is 9.

3. a.

Sample answer: The number of circles is odd and increases by 2 every time.

b. 11; 19

c. Sample answer: Since each step is the next odd number, I skip counted from 1 by 2 until I got to the 10th step.

5. 250,004

Exploring Square Numbers

A **square number** is a number that can be written as the product of a number multiplied by itself. For example, the square number 9 can be written as 3 * 3.

3 * 3 = 9

SRB
58-61

① Fill in the missing factors and square numbers.

Factors	Square Number
	4
3 * 3	9
4 * 4	
	25
	36

② What pattern(s) do you see in the factors? In the products?

_____

_____

③ What other pattern(s) do you see in the table?

_____

_____

④ Write an equation to describe each array.

a.

b.

Equation: _____   Equation: _____

⑤ a. Which of the arrays above shows a square number? _____

b. Explain. _____

Practice

⑥ 32, 45, 58, _____, _____, _____   Rule: _____

⑦ _____, _____, _____, 89, 115, 141   Rule: _____

Area of a Rectangle

① Draw a rectangle that has length of 9 units and width of 4 units.

② Draw a rectangle that has a length of 7 units and a width of 8 units.

SRB
202-204

Equation: _____

Area = _____ square units

Equation: _____

Area = _____ square units

Use the formula $A = l * w$ to find the area of each rectangle.

③
8'

6'

Equation: _____

Area = _____ square feet

④
3"

7"

Equation: _____

Area = _____ square inches

⑤ Riley's dining room tabletop is 9 feet long and 6 feet wide. What is the area of the tabletop?

Equation: _____

Area = _____ square feet

Practice

⑥ 368 − 59 = _____

⑦ 194 − 147 = _____

⑧ _____ = 1,729 − 623

39

Working with Factor Pairs

① Write equations to help you find all the factor pairs of each number below. Use dot arrays, if needed.

Number	Equations with Two Factors	Factor Pairs
6	1 * 6 = 6 2 * 3 = 6 3 * 2 = 6 6 * 1 = 6	1 and 6 2 and 3
9		
10		
17		
40		

Practice

② 356 + 433 = _____   ③ _____ = 2,167 + 696

④ _____ = 4,578 − 2,232   ⑤ 3,271 − 1,089 = _____

Finding Multiples

① List the first 5 multiples of 4. _____

② List the first 10 multiples of 2. _____

③ a. List the first 10 multiples of 3. _____

b. List the first 10 multiples of 5. _____

c. List the multiples of 3 that are also multiples of 5. _____

④ Is 28 a multiple of 7? _____ Explain. _____

_____

⑤ Is 35 a multiple of 6? _____ Explain. _____

_____

⑥ a. List the factors of 15. List the multiples through 15 of each factor.

Factors of 15	Multiples of the Factors (of 15)

b. Is 15 a multiple of each of its factors? _____ Explain. _____

_____

Practice

⑦ 24, _____, 48, _____, 72, _____   Rule: _____

⑧ _____, 108, 162, _____, 270, _____   Rule: _____

⑨ 86, _____, 52, _____, 18, _____   Rule: _____

⑩ 425, _____, 339, _____, 253, _____   Rule: _____

Prime and Composite Numbers

A **prime number** is a whole number that has exactly two different factors—1 and the number itself. A **composite number** is a whole number that has more than two different factors.

For each number:

- List all of its factors.

- Write whether the number is prime or composite.

- Circle all of the factors that are prime numbers.

	Number	Factors	Prime or Composite?
①	11		
②	19		
③	24		
④	29		
⑤	36		
⑥	49		
⑦	50		
⑧	70		
⑨	100		

Practice

Solve.

⑩ $841 + 527 =$ _____

⑪ _____ $= 3,263 + 5,059$

⑫ $7,461 + 2,398 =$ _____

⑬ _____ $= 4,172 - 3,236$

⑭ $8,158 = 5,071 +$ _____

⑮ $3,742 - 3,349 =$ _____

Using Multiplication

Home Market sells 3 grapefruits for $2.

SRB
42, 53

① Darius spent $6 on grapefruits. How many did he buy? Use words, numbers, or diagrams to show your reasoning.

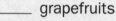 grapefruits

② Jana bought 15 grapefruits. How much did she spend? Use words, numbers, or diagrams to show your reasoning.

_____ dollars

③ On the back of this page, write a multiplication number story about buying grapefruits at Home Market. Show how to solve your number story.

Practice

Write these numbers using words.

④ 12,309 _____

⑤ 30,041 _____

⑥ 600,780 _____

⑦ 9,090,506 _____

47

Converting Units of Time

Use the measurement scales to fill in the tables and answer the questions.

SRB
198-199

minutes

```
0      1      2      3      4      5      6      7      8      9      10
|------|------|------|------|------|------|------|------|------|------|-->
0      60    120    180    240    300    360    420    480    540    600
```
seconds

hours

```
0      1      2      3      4      5      6      7      8      9      10
|------|------|------|------|------|------|------|------|------|------|-->
0  30  60  90 120 150 180 210 240 270 300 330 360 390 420 450 480 510 540 570 600
```
minutes

①
Hours	Minutes
1	60
4	
8	
11	

②
Minutes	Seconds
1	60
5	
10	
20	

③ Zac worked on his spelling for 9 minutes last night and 8 minutes this afternoon. How many seconds did he work? Answer: _____ seconds

④ Eton's baby sister took a nap for 2 hours and 22 minutes yesterday and 1 hour and 35 minutes today. How many more minutes did she sleep yesterday than today? Answer: _____ minutes

Try This

⑤ How many seconds did Eton's baby sister sleep all together?
Answer: _____ seconds

Practice

⑥ 945 + 1,055 = _____

⑦ 2,953 + 4,471 = _____

⑧ 4,552 + 4,548 = _____

⑨ 3,649 + 3,649 = _____

49

Multiplicative Comparisons

> **Family Note** In this lesson students used comparison statements and equations to represent situations in which one quantity is a number of times as much as another quantity. For example: José saved $5 over the summer. His sister saved 3 times as much. How much money did José's sister save? In this number story students compare the amount of money José saved to the amount his sister saved. Students write the equation $3 * 5 = 15$ to represent this comparison and solve the problem: José's sister saved $15. Because these comparison statements and equations involve multiplication, they are called multiplicative comparisons.

Complete the problems below. Write an equation with a letter for the unknown and solve.

SRB 56-57

① What number is 7 times as much as 9?

Equation with unknown:

_____

Answer: _____

② What number is 5 times as much as 6?

Equation with unknown:

_____

Answer: _____

③ 32 is 4 times as much as what number?

a. Equation with unknown: _____

b. Answer: _____

④ Write an equation to represent this situation and solve.

Ameer worked 3 times as many hours as Simi each week during the summer. If Simi worked 10 hours each week, how many hours did Ameer work each week?

a. Equation with unknown: _____

b. Answer: _____ hours

Practice

⑤ $7,482 - 7,083 =$ _____

⑥ $7,702 - 3,581 =$ _____

⑦ $5,201 - 3,052 =$ _____

⑧ $8,002 - 5,403 =$ _____

Solving Multiplicative Comparison Number Stories

Make a diagram or drawing and write an equation to represent the situation. Then find the answer.

① Judith collected 9 marbles. Swen has 6 times as many. How many marbles does Swen have?

Diagram or drawing:

Equation with unknown: _____

Answer: _____ marbles

② Sol ran 4 times as many minutes as Jerry. Jerry ran 12 minutes. How many minutes did Sol run?

Diagram or drawing:

Equation with unknown: _____

Answer: _____ minutes

Insert quantities into the number story. Make a diagram and write an equation to represent the story.

③ Lola picked _____ apples. Eilene picked _____ apples. Eilene picked _____ times as many apples as Lola.

Diagram or drawing:

Equation with unknown: _____

Answer: _____ apples

Practice
Write these numbers in expanded form.

④ 3,830 _____

⑤ 56,037 _____

⑥ 800,700 _____

⑦ 716,305 _____

Identifying Triangles

SRB
229, 233

Write the letter or letters that match each statement.

(1) Has perpendicular line segments _____

(2) Has an obtuse angle _____

(3) Has right angles _____

(4) Has acute angles _____

(5) Has more than one kind of angle _____

(6) Has only one kind of angle _____

(7) Does NOT have any right angles _____

(8) Is a right triangle _____

Practice

(9) List all the factors of 12. _____

(10) Name the next 4 multiples of 7. 35, _____, _____, _____, _____

Drawing Quadrilaterals

① A parallelogram is a quadrilateral that has 2 pairs of parallel sides. Draw a parallelogram.

SRB
234-235

② Answer each question, drawing pictures on the back of this page to help you.

a. Can a parallelogram have right angles? _____ Explain.

_____

_____

b. Could a quadrilateral have 4 obtuse angles? _____ Explain.

_____

_____

c. Name a quadrilateral that has at least 1 pair of parallel sides.

_____

③ Draw a quadrilateral that has at least 1 right angle.

④ Draw a quadrilateral that has 2 separate pairs of equal length sides but is NOT a parallelogram.

This is called a _____.

Practice

⑤ 5 * 30 = _____

⑥ _____ = 40 * 3

⑦ _____ = 80 * 6

⑧ 6 * 70 = _____

57

Drawing Lines of Symmetry

① Draw the other half of each picture to make it symmetrical.
Use a straightedge to form the line of symmetry.

② Draw a line of symmetry for each figure.

③ List four items in your home that are symmetric. Pick one item and draw it below, including at least one line of symmetry.

Item: _____   Item: _____   Drawing:

Item: _____   Item: _____

Practice

④ _____ = 2,767 + 3,254   ⑤ 193 + 6,978 = _____

⑥ 7,652 − 5,388 = _____   ⑦ _____ = 4,273 − 1,678

59

Identifying Patterns

① Complete.

in
↓

Rule

*9

↓
out

in	out
2	18
	27
4	
	45
6	

What patterns do you see?

_____

_____

_____

② Complete.

SRB
58-60,
65-67

in
↓

Rule

+ 11

↓
out

in	out
11	22
	33
33	
	55
55	

What patterns do you see?

_____

_____

_____

③ Study the pattern.

  1   2   3   4   5

a. Draw the next step in the pattern. What patterns do you notice?

_____

_____

b. How many circles will be in the 6th step? _____ In the 10th step? _____

c. How did you figure out how many circles will be in the 10th step?

_____

_____

Practice

④ 800,000 + 90 = _____

⑤ 200,000 + 50,000 + 4 = _____

61

Fractions and Decimals

Of the different types of numbers that elementary school students are required to understand, fractions can be the hardest. To succeed with fractions, students must build on their understanding of whole numbers, but they also have to understand how fractions differ from whole numbers.

This unit focuses on three big ideas about fractions: recognizing equivalent fractions, comparing fractions, and representing or showing fractions in different ways.

Fraction Equivalence

Equivalence, or equality, is one of the biggest ideas in mathematics. Much of arithmetic, for example, is really just rewriting numbers in equivalent forms. When we ask students to solve $850 + 125$, we are asking for a single number, 975, that is equivalent to $850 + 125$.

Students were introduced to equivalent fractions in third grade. They reasoned about equivalent fractions by thinking about sharing ($\frac{1}{2}$ is a fair share when 1 whole is shared 2 ways), division ($\frac{3}{4}$ is the result of dividing 3 wholes into 4 parts), and measurement ($\frac{1}{2}$ of an inch and $\frac{2}{4}$ of an inch name the same length).

In fourth grade students learn a multiplication rule for making equivalent fractions: *To make a fraction equivalent to a given fraction, multiply the numerator and denominator by the same number (so long as that number is not 0).* For example, to make a fraction equivalent to $\frac{3}{4}$, we can multiply the numerator and denominator by 2: $\frac{3*2}{4*2} = \frac{6}{8}$.

Comparing Fractions

When students compare fractions with the same denominator or numerator, they are also building on previous work by thinking about how familiar things are divided and shared. In third grade students reasoned that $\frac{3}{5}$ is less than $\frac{3}{4}$ because sharing 3 pizzas among 5 people means less pizza for each person compared to sharing the same 3 pizzas among only 4 people.

In fourth grade students continue work with fractions based on visual models and reasoning about sharing. They also learn to use benchmark fractions, such as 0, $\frac{1}{2}$, and 1. Students reason that $\frac{3}{5}$ is more than $\frac{1}{3}$ because $\frac{3}{5}$ is more than $\frac{1}{2}$ and $\frac{1}{3}$ is less than $\frac{1}{2}$.

Representing Fractions

One reason fractions can be hard to understand is that the same number can be shown in so many different ways. For example, $\frac{1}{2}$ can be written as $\frac{2}{4}$, $\frac{3}{6}$, or an infinite number of equivalent fractions. It can also be written as 0.5, 0.50, or an infinite number of other equivalent decimals. And the number $\frac{1}{2}$ can represent half the amount of different whole units.

Different representations of $\frac{1}{2}$

In this unit students work with fractions and decimals represented in many different ways, using concrete objects like fraction circle pieces, base-10 blocks, and strips of paper; hundred grids and number lines; drawings of circles and rectangles; and rulers and other measuring tools. For example, they might show that $\frac{3}{4}$ is greater than $\frac{5}{8}$ using fraction circle pieces, folding paper strips, pointing to marks on a ruler, or finding equivalent fractions with a common denominator or common numerator.

In this way students build a network of ideas that help them develop a solid conceptual understanding of fractions and decimals, which will support work in later units and grades that is focused on more formal rules and procedures.

Please keep this Family Letter for reference as your child works through Unit 3.

Vocabulary

Important terms in Unit 3:

benchmark A count or measure that can be used to evaluate the reasonableness of other counts, measures, or estimates. For example, a benchmark for land area is a football field, which is about 1 acre.

centimeter A metric unit of length equivalent to 10 millimeters, $\frac{1}{10}$ of a decimeter, and $\frac{1}{100}$ of a meter.

common denominator A nonzero number that is a multiple of the denominators of two or more fractions. For example, the fractions $\frac{1}{2}$ and $\frac{2}{3}$ have common denominators 6, 12, 18, and so on.

common numerator A nonzero number that is a multiple of the numerators of two or more fractions. For example, the fractions $\frac{3}{4}$ and $\frac{4}{5}$ have common numerators 12, 24, 36, and so on.

denominator The nonzero digit b in a fraction $\frac{a}{b}$. In a part-whole fraction, the denominator is the number of equal parts into which the whole has been divided.

equivalent Equal in value, but possibly represented in a different form. For example, $\frac{1}{2}$, 0.5, and 50% are all equivalent.

Equivalent Fractions Rule A rule stating that if the numerator and denominator of a fraction are multiplied by the same nonzero number, the result is a fraction that is equivalent to the original fraction. This rule can be represented as: $\frac{a}{b} = \frac{(n * a)}{(n * b)}$.

hundredth A single part out of one hundred equal parts that form a whole.

interval The points and their coordinates on a segment of a number line. The interval between 0 and 1 on a number line is the *unit interval*.

meter The basic metric unit of *length* from which other metric units of length are derived. One meter is equal to 10 decimeters, 100 centimeters, or 1,000 millimeters. A meter is a little longer than a yard.

metric The measurement system used in most countries and by virtually all scientists around the world. Units in the metric system are related by powers of 10.

millimeter A metric unit of length equal to $\frac{1}{10}$ of a centimeter or $\frac{1}{1,000}$ of a meter.

numerator The digit a in a fraction $\frac{a}{b}$. In a part-whole fraction in which the whole is divided into a number of equal parts, the numerator is the number of equal parts being considered.

reasoning An explanation or justification for how to solve a problem or answer a question.

representation Something that shows, symbolizes, or stands for something else. For example, numbers can be represented using base-10 blocks, spoken words, or written numerals.

strategy A general approach to solving a problem or answering a question.

tenth A single part out of ten equal parts that form a whole.

unit A label used to put a number in context. In measuring length, for example, inches and centimeters are units. In a problem about 5 apples, the unit is apples. In *Everyday Mathematics* students keep track of units in *unit boxes*.

whole An entire object, collection of objects, or quantity being considered in a problem situation; 100%.

Do-Anytime Activities

To work with your child on concepts taught in this unit, try these activities:

1. Have your child look for everyday uses of fractions in grocery stores, shoe sizes, cookbooks, measuring cups, and statistics in newspapers and on television.

2. Encourage your child to express fractions, quantities, and measures, such as a quarter of an hour, a quart of orange juice, or a quarter cup of milk.

3. Encourage your child to incorporate terms such as *whole, halves, thirds*, and *fourths* into his or her everyday vocabulary.

Building Skills through Games

In this unit your child will play the following games to develop his or her understanding of fractions and decimals. For detailed instructions, see the *Student Reference Book.*

Fraction Match See *Student Reference Book*, page 263. This game is for 2 to 4 players and requires one set of fraction cards. The game develops skill in naming equivalent fractions.

Fraction Top-It See *Student Reference Book*, page 265. This is a game for 2 to 4 players and requires one set of fraction cards. The game develops skill in comparing fractions.

Decimal Top-It See *Student Reference Book*, page 253. This is a game for 2 to 4 players and requires one set of number cards and a gameboard. The game develops skill in comparing decimals.

As You Help Your Child with Homework

As your child brings assignments home, you may want to go over instructions together, clarifying them as necessary. The answers listed below will guide you through the Home Links for this unit.

Home Link 3-1

1. $1\frac{1}{4}$, or $\frac{5}{4}$ pizzas

   One way:

   Another way:

3. 14, 21, 28, 35   5. 2, 3

Home Link 3-2

1.   3.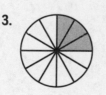

6. 25, 30, 35, 40   8. 4, 6, 8, 12, 16, 24, 48

Home Link 3-3

1. a. $\neq$ b. $=$ c. $\neq$ d. $=$ e. $=$

3. *c, d,* and *e* should be circled

5. 8,033   7. 288

Home Link 3-4

1. 12   3. 6   5. 2   7. 12   9. 3

11. Sample answers: $\frac{2}{4}$, $\frac{4}{8}$, $\frac{5}{10}$, $\frac{6}{12}$

13. 1 and 30, 2 and 15, 3 and 10, 5 and 6

Home Link 3-5

1. a. $\frac{3}{4}$ b. Answers vary.

3. They will get the same amount; $\frac{3}{4}$ and $\frac{6}{8}$ are equivalent fractions.

6. 1 and 75, 3 and 25, 5 and 15

Home Link 3-6

1. They have both read the same amount; Sample answer: $\frac{3}{4}$ is the same as $\frac{6}{8}$.

2. Heather; $\frac{5}{8} > \frac{5}{10}$, or $\frac{5}{10} < \frac{5}{8}$

3. Howard's; Sample answers: Jermaine's team won less than half of its games, and Howard's team won more than half; $\frac{2}{5}$ is the same as $\frac{4}{10}$, and $\frac{7}{10} > \frac{4}{10}$.

5. F   7. T

Home Link 3-7

1. $\frac{2}{6}$, $\frac{4}{6}$, $\frac{5}{6}$

3. $\frac{1}{6}$, $\frac{4}{10}$, $\frac{1}{2}$, $\frac{7}{12}$, $\frac{2}{3}$

5. 10,121   7. 1,329

Home Link 3-8

1.

Number in Words	Fraction	Decimal
one-tenth	$\frac{1}{10}$	0.1
four-tenths	$\frac{4}{10}$	0.4
eight-tenths	$\frac{8}{10}$	0.8
nine-tenths	$\frac{9}{10}$	0.9
two-tenths	$\frac{2}{10}$	0.2
seven-tenths	$\frac{7}{10}$	0.7

3. 0.5   **5.** 0.9

7. 1, 2, 4, 5, 10, 20, 25, 50, 100

9. 1 and 42, 2 and 21, 3 and 14, 6 and 7

Home Link 3-9

1. $\frac{30}{100}$; 0.30   **3.** $\frac{65}{100}$; 0.65

5.   **7.** 3, 9

Home Link 3-10

1. 57; 5; 7   **3.** 4; 0; 4   **5.** 8.4   **7.** 0.05

9. 0.04, 0.05, 0.06, 0.07, 0.08, 0.09

11. 46,000

Home Link 3-11

1. 0.02; 0.03; 0.04; 0.05; 0.06; 0.07

3–6.

7. **a.** > **b.** < **c.** >   **9.** 13,931   **11.** 1,569

Home Link 3-12

1. Answers vary.   **3.** 100; 1.8; 2,360; 572; 65

5. 1 and 60, 2 and 30, 3 and 20, 4 and 15, 5 and 12, 6 and 10

Home Link 3-13

1. <   **3.** >   **5.** >   **7.** =

9. hundredths; 0.09   **11.** 6.59, 6.60, 6.61

13. 4.4   **15.** 8.1

17. 56,230   **19.** 15,379

Sharing Equally

Use drawings to help you solve the problems. Solve each problem in more than one way. Show your work.

① Four friends shared 5 pizzas equally. How much pizza did each friend get?

_____ pizzas

One way:

Another way:

② Five kittens are sharing 6 cups of milk equally. How much milk does each kitten get?

_____ cups of milk

One way:

Another way:

Practice

③ Name the next 4 multiples of 7. 7, _____, _____, _____, _____

④ List all the factors of 18. _____

⑤ List all the factors of 18 that are prime. _____

⑥ List all the factor pairs of 40.

_____ and _____ ; _____ and _____ ;

_____ and _____ ; _____ and _____

Fraction Circles

① Divide into 4 equal parts. Shade $\frac{1}{4}$.

② Divide into 8 equal parts. Shade $\frac{2}{8}$.

SRB
136

③ Divide into 12 equal parts. Shade $\frac{3}{12}$.

④ Create your own. Divide into equal parts and shade a portion. Record the amount you shaded.

⑤ What patterns do you notice in Problems 1 through 3?

_____

_____

Practice

⑥ List the next 4 multiples of 5. 20, _____, _____, _____, _____

⑦ List all the factors of 48. _____

⑧ List the factors of 48 that are composite. _____

Finding Equivalent Fractions

Use the number lines to help you answer the following questions.

(1) Fill in the blank with = or ≠.

a. $\frac{2}{3}$ ≠ $\frac{1}{3}$

b. $\frac{2}{6}$ ≠ $\frac{1}{3}$

c. $\frac{2}{6}$ ≠ $\frac{2}{5}$

d. $\frac{1}{5}$ = $\frac{2}{10}$

e. $\frac{2}{12}$ = $\frac{1}{6}$

(2) Fill in the missing numbers.

a. $\frac{1}{5} = \frac{\boxed{2}}{10}$

b. $\frac{4}{12} = \frac{\boxed{1}}{3}$

c. $\frac{5}{10} = \frac{\boxed{1}}{2}$

d. $\frac{3}{6} = \frac{\boxed{6}}{12}$

e. $\frac{4}{6} = \frac{\boxed{2}}{3}$

(3) Circle the number sentences that are NOT true.

a. $\frac{3}{12} = \frac{1}{4}$ ⟵ circled

b. $\frac{1}{2} = \frac{5}{10}$ ⟵ circled

c. $\frac{2}{6} = \frac{2}{5}$ ⟵ circled

d. $\frac{7}{10} = \frac{4}{6}$

e. $\frac{9}{10} = \frac{11}{12}$

Practice

Solve using U.S. traditional addition or subtraction.

(4) _____ = 989 + 657

(5) 3,314 + 4,719 = _____

(6) 5,887 − 3,598 = _____

(7) _____ = 2,004 − 1,716

Finding Equivalent Fractions

Family Note Today students learned about an **Equivalent Fractions Rule**, which can be used to rename any fraction as an equivalent fraction. The rule for multiplication states that if the numerator and denominator are multiplied by the same nonzero number, the result is a fraction that is equivalent to the original fraction.

For example, the fraction $\frac{1}{2}$ can be renamed as an infinite number of equivalent fractions. When you multiply the numerator 1 by 5, the result is 5. When you multiply the denominator 2 by 5, the result is 10.

$$\frac{1 \times 5}{2 \times 5} = \frac{5}{10}$$

This results in the number sentence $\frac{1}{2} = \frac{5}{10}$. If you multiplied both the numerator and denominator in $\frac{1}{2}$ by 3, the result would be $\frac{3}{6}$, which is also equal to $\frac{1}{2}$.

Fill in the boxes to complete the equivalent fractions.

Example: $\frac{1}{2} = \frac{3}{\boxed{6}}$

① $\frac{1}{2} = \dfrac{6}{\boxed{}}$ ② $\frac{1}{4} = \dfrac{3}{\boxed{}}$ ③ $\frac{1}{3} = \dfrac{2}{\boxed{}}$ ④ $\frac{2}{3} = \dfrac{8}{\boxed{}}$ ⑤ $\frac{1}{5} = \dfrac{\boxed{}}{10}$

⑥ $\frac{2}{5} = \dfrac{\boxed{}}{10}$ ⑦ $\frac{3}{4} = \dfrac{9}{\boxed{}}$ ⑧ $\frac{5}{6} = \dfrac{10}{\boxed{}}$ ⑨ $\frac{2}{\boxed{}} = \dfrac{6}{9}$ ⑩ $\frac{4}{\boxed{}} = \dfrac{8}{12}$

⑪ Name 3 equivalent fractions for $\frac{1}{2}$. _____

Practice

⑫ List all the factors of 56. _____

⑬ Write the factor pairs for 30.

_____ and _____, _____ and _____, _____ and _____,

_____ and _____

⑭ Is 30 prime or composite? _____

SRB
141

Sharing Veggie Pizza

① Karen and her 3 friends want to share 3 small veggie pizzas equally. Karen tried to figure out how much pizza each of the 4 children would get. She drew this picture and wrote two answers.

$\frac{3}{4}$ or $1\frac{1}{4}$

a. Which of Karen's answers is correct? $\frac{3}{4}$

b. Draw on Karen's diagram to make it clear how the pizza should be distributed among the 4 children.

② Erin and her 7 friends want to share 6 small veggie pizzas equally.

How much pizza will each of the 8 children get? 6/8

③ Who will get more pizza, Karen or Erin? They got the same amount.

Explain or show how you know. Using the equivalent fractions rule, he know that $\frac{3}{4}$

Practice

④ List all the factors of 50. 1 & 50, 5 & 10, 2 & 25

⑤ Is 50 prime or composite? composite

⑥ Write the factor pairs for 75.

1 and 75

3 and 25

5 and 15

Solving Fraction Comparison Number Stories

Solve the problems below.

SRB
145-146

(1) Tenisha and Christa were each reading the same book. Tenisha said she was $\frac{3}{4}$ of the way done with it, and Christa said she was $\frac{6}{8}$ of the way finished.

Who has read more, or have they read the same amount? _____

_____

How do you know? _____

_____

(2) Heather and Jerry each bought an ice cream bar. Although the bars were the same size, they were different flavors. Heather ate $\frac{5}{8}$ of her ice cream bar, and Jerry ate $\frac{5}{10}$ of his.

Who ate more, or did they eat the same amount? _____

Write a number sentence to show this. _____

(3) Howard's baseball team won $\frac{7}{10}$ of its games. Jermaine's team won $\frac{2}{5}$ of its games. They both played the same number of games.

Whose team won more games, or did they win the same amount? _____

How do you know? _____

_____

(4) Write your own fraction number story. Ask someone at home to solve it.

_____

_____

_____

Practice

Write T for true or F for false.

(5) $1,286 + 2,286 = 3,752$ ____

(6) $9,907 - 9,709 = 200$ ____

(7) $2,641 + 4,359 = 2,359 + 4,641$ ____

(8) $2,345 - 198 = 2,969 - 822$ ____

Comparing and Ordering Fractions

Write the fractions from smallest to largest, and then justify your conclusions by placing the numbers in the correct places on the number lines.

SRB
135,
147-148

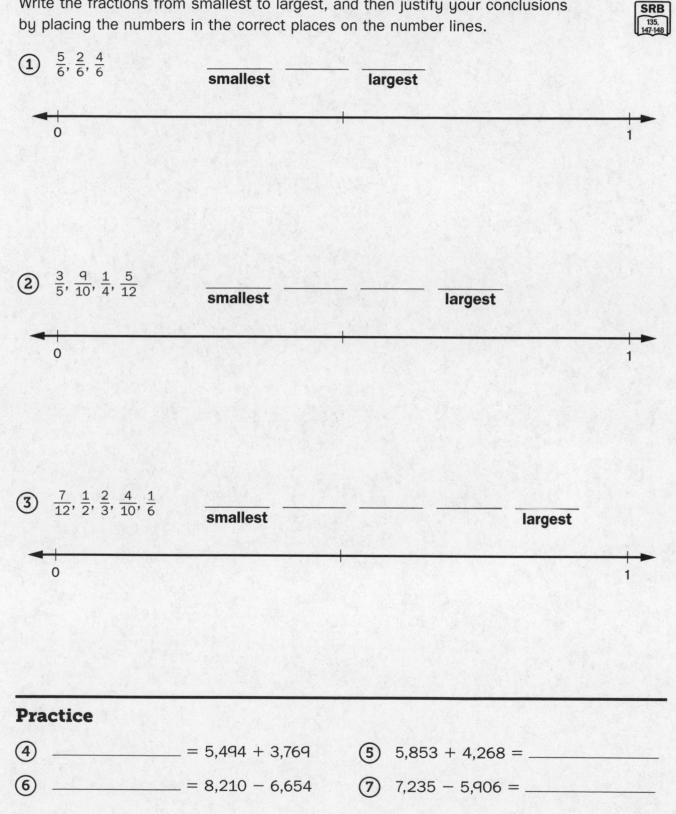

(1) $\frac{5}{6}, \frac{2}{6}, \frac{4}{6}$

_____ _____ _____
smallest   **largest**

0   1

(2) $\frac{3}{5}, \frac{9}{10}, \frac{1}{4}, \frac{5}{12}$

_____ _____ _____ _____
smallest   **largest**

0   1

(3) $\frac{7}{12}, \frac{1}{2}, \frac{2}{3}, \frac{4}{10}, \frac{1}{6}$

_____ _____ _____ _____ _____
smallest   **largest**

0   1

Practice

(4) _____ = 5,494 + 3,769   (5) 5,853 + 4,268 = _____

(6) _____ = 8,210 − 6,654   (7) 7,235 − 5,906 = _____

Names for Fractions and Decimals

(1) Fill in the blanks in the table below.

SRB
150-151

Number in Words	Fraction	Decimal
one-tenth		
four-tenths		
	$\frac{8}{10}$	
		0.9
	$\frac{2}{10}$	
seven-tenths		

(2) Name two ways you might see decimals used outside of school.

_____

```
        M                         P
├──┼───────────┼──────────────────┼──────┼──►
0                                         1
```

(3) What decimal is represented by the tick mark labeled *M*? _____

(4) What fraction is represented by the tick mark labeled *M*? _____

(5) What decimal is represented by the tick mark labeled *P*? _____

(6) What fraction is represented by the tick mark labeled *P*? _____

Practice

(7) List all the factors of 100. _____

(8) List the factors of 100 that are prime. _____

(9) Write the factor pairs for 42.

_____ and _____   _____ and _____

_____ and _____   _____ and _____

83

Representing Fractions and Decimals

If the grid is the whole, then what part of each grid is shaded?

Write a fraction and a decimal below each grid.

SRB
150-151

①

fraction: _____

decimal: _____

②

fraction: _____

decimal: _____

③

fraction: _____

decimal: _____

④ Color 0.8 of the grid.

⑤ Color 0.04 of the grid.

⑥ Color 0.53 of the grid.

Practice

⑦ The numbers 81, 27, and 45 are all multiples of 1, _____, and _____.

⑧ List the first ten multiples of 6.

_____, _____, _____, _____, _____, _____, _____, _____,

_____, _____

Tenths and Hundredths

Family Note Your child continues to work with decimals. Encourage him or her to think about ways to write money amounts. This is called dollars-and-cents notation. For example, $0.07 (7 cents), $0.09 (9 cents), and so on.

Write the decimal numbers that represent the shaded part in each diagram.

Whole

grid

SRB
149-150

①   _____ hundredths

_____ tenths _____ hundredths

②   _____ hundredths

_____ tenths _____ hundredths

③   _____ hundredths

_____ tenths _____ hundredths

Write the words as decimal numbers.

④ twenty-three hundredths

_____

⑤ eight and four-tenths

_____

⑥ thirty and twenty-hundredths

_____

⑦ five-hundredths

_____

Continue each pattern.

⑧ 0.1, 0.2, 0.3, _____, _____, _____, _____, _____, _____

⑨ 0.01, 0.02, 0.03, _____, _____, _____, _____, _____, _____

Practice

⑩ Round 7,604 to the nearest thousand. _____

⑪ Round 46,099 to the nearest thousand. _____

⑫ Round 8,500,976 three ways: nearest thousand, hundred-thousand, and million.

_____   _____   _____

87

Practice with Decimals

Fill in the missing numbers.

(1)

0   0.01   _____   _____   _____   _____   _____   _____   0.08

(2)

0.7   _____   _____   _____   _____   _____   _____   _____   1.5

Follow these directions on the ruler below.

SRB
154,
182-183

(3) Make a dot at 7 cm and label it with the letter *A*.

(4) Make a dot at 90 mm and label it with the letter *B*.

(5) Make a dot at 0.13 m and label it with the letter *C*.

(6) Make a dot at 0.06 m and label it with the letter *D*.

```
0   1   2   3   4   5   6   7   8   9   10  11  12  13  14  15
cm
```

(7) Write <, >, or =.

   a. 1.2 ____ 0.12   **b.** 0.3 ____ 0.38   **c.** 0.80 ____ 0.08

(8) Complete. $\boxed{1 \text{ cm} = 10 \text{ mm}}$ $\boxed{1 \text{ m} = 100 \text{ cm}}$

cm	m
100	1
	5
1,000	
6,000	

cm	m
1	*0.01*
	0.03
	0.06
40	

Practice

(9) 6,366 + 7,565 = _____   (10) 3,238 + 29,784 = _____

(11) 9,325 − 7,756 = _____   (12) 14,805 − 2,927 = _____

Measuring Centimeters and Millimeters

(1) Find 6 objects in your home to measure. Use the ruler from the bottom of the page to measure them, first in centimeters and then in millimeters. Record your objects and their measurements.

SRB
180,
182-183

Example: ___*crayon*___ ___*3.5*___ cm ___*35*___ mm

Object   **Object**

_____ ____ cm ____ mm   _____ ____ cm ____ mm

_____ ____ cm ____ mm   _____ ____ cm ____ mm

_____ ____ cm ____ mm   _____ ____ cm ____ mm

Fill in the tables.

(2)

cm	mm
1	
15	
3.7	
49.6	
0.8	

(3)

cm	m
	1
180	
	23.6
	5.72
	0.65

Practice

(4) List the factors for 63. _____

(5) Write the factor pairs for 60.

_____ and _____   _____ and _____   _____ and _____

_____ and _____   _____ and _____   _____ and _____

cm	0	1	2	3	4	5	6	7	8	9	10	11	12	13	14	15
		.5	1.5	2.5	3.5	4.5	5.5	6.5	7.5	8.5	9.5	10.5	11.5	12.5	13.5	14.5

Fold

| | | 5 | 15 | 25 | 35 | 45 | 55 | 65 | 75 | 85 | 95 | 105 | 115 | 125 | 135 | 145 |
| mm | 0 | 10 | 20 | 30 | 40 | 50 | 60 | 70 | 80 | 90 | 100 | 110 | 120 | 130 | 140 | 150 |

91

Comparing Decimals

> **Family Note** Ask your child to read the decimal numerals aloud. Encourage your child to use the following method:
> 1. Read the whole-number part.
> 2. Say *and* for the decimal point.
> 3. Read the digits after the decimal point as though they form their own number.
> 4. Say *tenths* or *hundredths*, depending on the placement of the right-hand digit. Encourage your child to exaggerate the *-ths* sound. For example, 2.37 is read as "two and thirty-seven hundredths."

Write >, <, or =.

SRB
154-155

(1) 2.35 _____ 2.57   (2) 1.08 _____ 1.8

(3) 0.64 _____ 0.46   (4) 0.90 _____ 0.9

(5) 42.1 _____ 42.09   (6) 7.09 _____ 7.54

(7) 0.4 _____ 0.40   (8) 0.26 _____ 0.21

> means *is greater than*

< means *is less than*

Example: The 4 in 0.47 stands for 4 __*tenths*__ or __*0.4*__.

(9) The 9 in 4.59 stands for 9 _____ or _____.

(10) The 3 in 3.62 stands for 3 _____ or _____.

Continue each number pattern.

(11) 6.56, 6.57, 6.58, _____, _____, _____

(12) 0.73, 0.83, 0.93, _____, _____, _____

Write the number that is 0.1 more.   Write the number that is 0.1 less.

(13) 4.3 _____   (14) 4.07 _____   (15) 8.2 _____   (16) 5.63 _____

Practice

(17) 43,589 + 12,641 = _____   (18) 63,274 + 97,047 = _____

(19) 41,805 − 26,426 = _____   (20) 82,004 − 11,534 = _____

93

Multidigit Multiplication

In Unit 4 your child will multiply multidigit numbers using **extended multiplication facts,** **partial-products multiplication,** and **lattice multiplication**. Throughout the unit, students use these methods to solve real-life multistep multiplication number stories.

The unit begins with extended multiplication facts. Knowing that $5 * 3 = 15$ helps students see that $50 * 3 = 150$; $500 * 3 = 1,500$; and so on. Working with extended facts gives students the ability to multiply larger numbers with ease.

Students also learn the partial-products multiplication method in which the value of each digit in one factor is multiplied by the value of each digit in the other factor. They partition a rectangle into smaller parts to help them understand how the method works. The example below shows how to use partial-products multiplication to find $456 * 4$.

Partitioned Rectangles	Partial-Products Multiplication

To practice multiplying 2-digit numbers using partial-products multiplication, students play a game called *Multiplication Wrestling*.

Finally, students are introduced to the lattice multiplication method: The lattice method breaks down the numbers into place values, allowing students to work with smaller numbers while solving a multidigit multiplication problem. It is an efficient method, often taking no more time than other methods.

In this unit, students apply their understanding of multidigit multiplication to solve conversion problems involving liters and milliliters and grams and kilograms. They also find the area of rectilinear figures.

Please keep this Family Letter for reference as your child works through Unit 4.

Vocabulary

Important terms in Unit 4:

adjacent Next to, or adjoining.

decompose To "break apart" numbers into friendlier numbers.

Distributive Property A rule saying that if a, b, and c are real numbers, then: $a * (b + c) = (a * b) + (a * c)$.

extended multiplication facts Multiplication facts involving multiples of 10, 100, and so on. For example, $400 * 6 = 2{,}400$ and $20 * 30 = 600$ are extended multiplication facts.

gram (g) A unit of mass in the metric system. There are about 454 grams in 1 pound.

kilogram (kg) 1,000 grams.

lattice multiplication A way to multiply multidigit numbers. *For example:*

liter (L) A unit of capacity in the metric system. It is equivalent to a little more than one quart.

mass The measure of the amount of matter in an object.

milliliter (mL) $\frac{1}{1000}$ of a liter.

partial-products multiplication A way to multiply in which the value of each digit in one factor is multiplied by the value of each digit in the other factor. The final product is the sum of the partial products. *For example:*

$$
\begin{array}{r}
7\ 3 \\
*\quad 4\ 6 \\
\hline
\end{array}
$$

$$
\begin{array}{rl}
40 * 70 \rightarrow & 2\ 8\ 0\ 0 \\
40 * 3\ \ \rightarrow & 1\ 2\ 0 \\
6 * 70 \rightarrow & 4\ 2\ 0 \\
6 * 3 \rightarrow & +\quad 1\ 8 \\
\hline
& 3,\ 3\ 5\ 8
\end{array}
$$

partition (in partial-products multiplication) A technique that uses the Distributive Property to break up a large rectangle into smaller rectangles in order to find the area more easily in parts.

rectilinear figure A single figure formed by combining multiple adjacent rectangles.

Do-Anytime Activities

To work with your child on concepts taught in this unit, try these activities:

1. Practice extended multiplication facts such as $50 * 40 = \underline{\hspace{2em}}$.

2. Collect three to five cans and bottles from the kitchen. Put them on the table and ask your child to order them, without looking at the labels, based on the amount of liquid each container can hold and/or their mass. Ask your child to estimate both. Check the results together by looking at the labels.

3. Pose a multiplication problem and ask your child to solve it using a method of his or her choice. Have your child explain to you or someone else at home what he or she did to complete the problem.

Building Skills through Games

In this unit your child will play the following game to develop his or her understanding of multiplication. For detailed instructions, see the *Student Reference Book*.

Multiplication Wrestling See *Student Reference Book*, page 267.
The game provides practice with multiplication of 2-digit numbers by 2-digit numbers.

As You Help Your Child with Homework

As your child brings assignments home, you may want to go over instructions together, clarifying them as necessary. The answers listed below will guide you through the Home Links for this unit.

Home Link 4-1

1. 560; 3,200; 630; 3,600

3. 450; 200; 63,000; 28,000

5. 9; 240; 700; 6,300

7. Answers vary.　　9. 1,190

11. 13,303

Home Link 4-2

Number models are sample answers.

1. $(20 * 30) - (10 * 30) = 300$;
   330; Answers vary.

3. $30 * 50 = 1,500$; $30 * 40 = 1,200$; $1,500 - 1,200 = 300$;　496; Answers vary.

5. 1,410,000

Home Link 4-3

1. 140; Sample answer:

3. 441; Sample answer:

5. 2,956　　　　7. 2,559

Home Link 4-4

1. 8,000; 15,000; 20,000; 25,000

3. 122,000 mL　　5. 14,445　　7. 62,341

Home Link 4-5

1. Sample answer: Four calculators fit in a layer. The box is 5 cm tall, so there are 5 layers of calculators. The box fits 4 calculators * 5, which is 20 calculators in all.

3. 108　　　　　5. 129

Home Link 4-6

1.
```
    4 8
  *   3
  ─────
  1 2 0
+ 2 4
─────
  1 4 4
```

3. 9 [100,000s] + 5 [1,000s] + 6 [100s] + 3 [1s]

5. 2 [1,000,000s] + 5 [100,000s] + 9 [10,000s] +
9 [1,000s] + 2 [1s]

Home Link 4-7

1. 25; 50,000; 75,000; 100

3. 237,000; 98,000; 485; 920,000

5. 63,000 grams   **7.** 396   **9.** 294

Home Link 4-8

1. $478   **3.** $55

5. 1, 3, 7, 21   **7.** 1, 2, 3, 4, 6, 9, 12, 18, 36

Home Link 4-9

1. 1,748

```
        4 6
   *    3 8
   ---------
    1 2 0 0
      1 8 0
      3 2 0
   +     4 8
   ---------
    1, 7 4 8
```

3. $65 * 22 = t$; 1,430 trees

5. 185   **7.** 1,992

Home Link 4-10

1. 42; 420; 420; 4,200; 4,200; 42,000

3. 32; 320; 320; 3,200; 3,200; 32,000

5. 6; 6; 60; 9; 900; 9,000

7. 2,139   **9.** 32,632

Home Link 4-11

1. $18 * 27 = 486$; 486 square units

3. Sample answer: $100 * 30 = 3,000$;
$20 * 20 = 400$; $3,000 + 400 = 3,400$;
3,400 square inches

5. 1, 2, 31, 62   **7.** 1, 5, 11, 55

Home Link 4-12

Sample number models:

1. $(10 * 7) * 2 = 140$; $(5 * 7) * 2 = 70$;
$140 + 70 = 210$ stickers;

$(8 * 7) * 2 = x$; $(5 * 7) * 2 = y$;
$112 + 70 = s$; 182 stickers

3. 1 and 50, 2 and 25, 5 and 10

5. 1 and 85, 5 and 17

Home Link 4-13

1. 536

2. 5,852

4. 2,552

6. 616   **8.** 356

Multiplication Puzzles

Solve the multiplication puzzles mentally. Fill in the blank boxes.

SRB 102

Examples:

*	300	2,000
2	600	4,000
3	900	6,000

*	80	50
4	320	200
8	640	400

①

*	70	400
8		
9		

②

*	5	7
80		
600		

③

*	9	4
50		
7,000		

④

*		600
7	3,500	
		2,400

⑤

*		8
30	270	
		5,600

⑥

*	400	
	3,600	
20		10,000

Make up and solve some puzzles of your own.

⑦

*		

⑧

*		

Practice

Solve using U.S. traditional addition or subtraction.

⑨ 321 + 869 = _____

⑩ 5,401 − 752 = _____

⑪ 4,568 + 8,735 = _____

⑫ 9,156 − 4,584 = _____

Finding Estimates and Evaluating Answers

Write an estimate and show your thinking. Solve using a calculator. Check to see that your answer is reasonable.

SRB
82-89

① Alice sleeps an average of 9 hours per night. A cat can sleep up to 20 hours per day. About how many more hours does a cat sleep in 1 month than Alice?

Estimate: _____

Answer: About _____ more hours per month

Is your answer reasonable? _____ How do you know? _____

_____

② Koalas sleep about 22 hours a day. Pandas sleep about 10 hours a day. About how many more hours does a typical koala sleep in 1 year than a typical panda?

Estimate: _____

Answer: About _____ more hours per year

Is your answer reasonable? _____ How do you know? _____

_____

③ There are 30 Major League Baseball (MLB) teams and 32 National Football League (NFL) teams. The expanded roster for MLB teams is 40 players and it is 53 for NFL teams. How many more players are in the NFL than in the MLB?

Estimate: _____

Answer: _____ more players

Is your answer reasonable? _____ How do you know? _____

_____

Practice

Round to the nearest thousand.

④ 45,493 _____

Round to the nearest ten-thousand.

⑤ 1,409,836 _____

Partitioning Rectangles

Solve the multiplication problems by partitioning a rectangle. Then add each part of the rectangle to get the product.

SRB
103-104

Example: 5 * 72 = __*360*__

① 4 * 35 = _____

```
        70        2        350
   ┌─────────┬──────┐     + 10
 5 │   350   │  10  │     ─────
   └─────────┴──────┘      360
        72
```

② 6 * 83 = _____

③ 9 * 49 = _____

Practice

Solve using U.S. traditional addition or subtraction.

④ 9,289 + 1,476 = _____

⑤ 6,503 − 3,547 = _____

⑥ 5,619 + 5,999 = _____

⑦ 5,005 − 2,446 = _____

Converting Liquid Measures

Complete the table.

SRB
193-194

①

Liters (L)	Milliliters (mL)
8	
15	
20	
25	

② Mrs. Wong's students kept track of how much water they used to water the classroom plants. The first week they used 24 liters, and the second week they used 17 liters. How many more milliliters did they use the first week than the second?

Answer: _____ mL

③ My fish tank holds 64 liters of water. My neighbor's tank holds 58 liters of water. How many milliliters is that combined?

Answer: _____ mL

④ Mrs. Reyes filled her kiddie pool with 83 liters of water. Her children added 2,000 mL of water to the pool. How many liters of water are in the pool now?

Answer: _____ L

Practice

Solve using U.S. traditional addition or subtraction.

⑤ $4,638 + 9,807 =$ _____

⑥ $7,322 - 3,741 =$ _____

⑦ $55,812 + 6,529 =$ _____

⑧ $98,001 - 7,443 =$ _____

Using Multiplication

Ms. Patel wants to keep her classroom calculators in a box that is 25 centimeters long, 15 centimeters wide, and 5 centimeters tall. The calculators measure 12 centimeters long, 7 centimeters wide, and 1 centimeter tall. How many calculators can Ms. Patel fit in the box?

SRB
103-105

5 cm   25 cm

15 cm

1 cm   12 cm

7 cm

(1) Solve this problem. Show or explain how you solved the problem.

(2) Show or explain how you know your answer makes sense.

Practice

Sketch a rectangle or use partial products to solve.

(3) 27 * 4 = _____

(4) 48 * 9 = _____

(5) 43 * 3 = _____

(6) 81 * 5 = _____

107

Multiplying in Parts

In the example, a rectangle was drawn to represent the problem. Then partial-products multiplication was used to record the work in a simpler way. Use partial-products multiplication to solve Problems 1 and 2.

Example:

Partitioned Rectangle

	400	30	2
7	2800	210	14

432

Partial-Products Multiplication

```
    4 3 2
  *     7
  2800
   210
+   14
  3,024
```

①
```
    4 8
  *   3
```

②
```
    6 5 3
  *     8
```

Practice

Write the numbers in expanded form.

③ 905,603 _____

④ 589,043 _____

⑤ 2,599,002 _____

⑥ 8,003,952 _____

Using a Measurement Scale

① Fill in the blanks on the measurement scale.

```
kg  0        _____        50        75        _____
    |--------|--------|--------|--------|--------|
g   0      25,000                            100,000
              _____   _____
```

Complete the two-column tables.

②

Kilograms (kg)	Grams (g)
6	
14	
	27,000
101	

③

Kilograms (kg)	Grams (g)
237	
98	
	485,000
920	

④ Find three items in your home that have the mass listed in grams or kilograms. Be sure to tell whether the mass is kilograms or grams.

Item	Mass in Kilograms (kg) or Grams (g)

⑤ Among other foods, a giraffe in a zoo eats 4 kg of plant pellets and 5 kg of hay each day. How many grams of these foods does a giraffe eat in one week?

Answer: _____ grams

Practice

⑥ 52 * 7 = _____

⑦ 99 * 4 = _____

⑧ 61 * 8 = _____

⑨ 49 * 6 = _____

SRB
188-189

Money Number Stories

Family Note Today your child solved multistep number stories involving multiplication, addition, and subtraction of money amounts. Have your child explain a plan for solving each of the following problems and then solve it.

Mr. Russo is buying equipment for his baseball team. Use the table to the right to answer questions about his purchases.

Item	Price
Wooden bat	$49
Metal bat	$74
Glove	$35
Helmet	$22

(1) Mr. Russo needs 9 helmets and 8 gloves. How much will they cost in all?

Answer: $_____

(2) Mr. Russo wants to buy 6 bats for his team. How much more would it cost for him to buy 6 metal bats than 6 wooden bats?

Answer: $_____

(3) Mr. Russo buys 5 wooden bats and gives the cashier $300. How much change does he get?

Answer: $_____

(4) If the cashier only has $10 and $1 bills, what are two ways he could make Mr. Russo's change?

Answer: _____

_____

Practice

List the factors for the following numbers:

(5) 21 _____ _____ _____ _____

(6) 40 _____ _____ _____ _____ _____

(7) 36 _____ _____ _____ _____ _____

(8) 45 _____ _____ _____ _____

Practicing Partial-Products Multiplication

Solve using partial-products multiplication.

①   46 * 38 = _____

②   6 5
       * 3 2
       ‾‾‾‾‾

③   Donnie and Raj went apple picking at an orchard that had 65 rows of trees. Each row had 22 trees in it. How many trees were in the orchard?

Number model with unknown: _____

Answer: _____ trees

④   A new apartment building has 33 floors, with 24 apartments on each floor. How many apartments are in the building?

Number model with unknown: _____

Answer: _____ apartments

Practice

⑤   37 * 5 = _____

⑥   27 * 6 = _____

⑦   332 * 6 = _____

⑧   2,958 * 7 = _____

Extended Multiplication Facts

Solve mentally.

SRB
56-57,
102

① 6 * 7 = _____

 6 * 70 = _____

 60 * 7 = _____

 60 * 70 = _____

 600 * 7 = _____

 60 * 700 = _____

② 5 * 6 = _____

 5 * 60 = _____

 50 * 6 = _____

 50 * 60 = _____

 500 * 6 = _____

 50 * 600 = _____

③ 4 * 8 = _____

 4 * 80 = _____

 40 * 8 = _____

 40 * 80 = _____

 400 * 8 = _____

 40 * 800 = _____

④ 5 * _____ = 15

 30 * _____ = 150

 30 * _____ = 1,500

 _____ * 50 = 150

 _____ * 500 = 1,500

 30 * _____ = 15,000

⑤ 54 is _____ times as many as 9.

 540 is _____ times as many as 90.

 5,400 is _____ times as many as 90.

 540 is 60 times as many as _____.

 5,400 is 6 times as many as _____.

 54,000 is 6 times as many as _____.

Practice

Solve using U.S. traditional addition or subtraction.

⑥ 6,419 + 7,809 = _____

⑦ 8,045 − 5,906 = _____

⑧ 76,543 + 84,086 = _____

⑨ 65,409 − 32,777 = _____

Finding the Area

(1) Find the area.

Equation: _____

Answer: _____ square units

27

18

(2) A tool bench is 35 inches long and 19 inches wide.
How many square inches of the basement floor does it cover?

Equation: _____

Answer: _____ square inches

(3) Find the area.

20 in.

20 in.

40 in.

40 in.

30 in.

30 in.

100 in.

Equations: _____

_____

Answer: _____ square inches

Practice

List all of the factors for the numbers below.

(4) 48 _____

(5) 62 _____

(6) 63 _____

(7) 55 _____

119

Multistep Multiplication Number Stories

Write estimates and number models for each problem. Then solve.

SRB
26,
36-37

(1) Rosalie is collecting stickers for a scrapbook. She collected 8 stickers per day for 2 weeks and then collected 5 stickers per day for 2 weeks. How many stickers has Rosalie collected?

Estimate: _____

Number models with unknowns:

_____

_____

Answer: _____ stickers

(2) Rashaad's sister gives him 2 packs of baseball cards per month. Each pack has 11 cards. She gives him 3 extra packs for his birthday. How many cards does Rashaad get in a year?

Estimate: _____

Number models with unknowns:

_____

_____

Answer: _____ cards

Does your answer make sense? Explain. _____

_____

_____

Practice

Name all the factor pairs.

(3) 50 _____

(4) 72 _____

(5) 85 _____

(6) 90 _____

Lattice Multiplication

Use the lattice method to find the products.

Example 5 * 46 = _**230**_

```
        4   6
      ┌───┬───┐
    2 │ 2/│ 3/│ 5
      │ /0│ /0│
      └/──┴/──┘
       3   0
```

(1) 8 * 67 = _____

(2) 7 * 836 = _____

(3) 6 * 531 = _____

(4) 44 * 58 = _____

(5) 84 * 78 = _____

Practice

(6) 77 * 8 = _____

(7) 49 * 2 = _____

(8) 89 * 4 = _____

(9) 183 * 5 = _____

123

Fraction and Mixed-Number Computation; Measurement

In Unit 3 students learned how to compare and order fractions and decimals. In Unit 5 they deepen their understanding by learning how a fraction such as $\frac{3}{4}$ can be broken into smaller parts, such as $\frac{1}{4} + \frac{1}{4} + \frac{1}{4}$. Based on this understanding, students are able to see how adding and subtracting fractions with like denominators is simply putting together or taking away some number of same-size parts. For example, $\frac{3}{4} - \frac{1}{4}$ can be thought of as taking away 1 of the 3 parts, or fourths, that make up the fraction $\frac{3}{4}$.

In this unit students extend this idea to adding and subtracting mixed numbers, such as $1\frac{1}{4} + 2\frac{2}{4}$. They use different fraction representations and tools, including fraction circles, number lines, and drawings, to build a concrete understanding of the meaning of fractions, as opposed to just learning rules and procedures.

Line Plots

Line plots are used to organize and display data. As you can see from the diagram, a line plot can be thought of as a rough sketch of a bar graph.

From this line plot we can learn that the tallest books are 9 inches tall and that there are 3 of them, that no books are $8\frac{3}{4}$ inches tall, and so on.

Students also create line plots with data they collect in fractional units and then use information in the plots to solve problems involving adding and subtracting fractions.

Book Heights

Height (in inches)

Angles: Unit Iteration and Rotations

Students begin their work with angle measurement by exploring the attribute of angle size. They begin measuring angles using a nonstandard unit—a wedge—as a way to see how measuring an angle is the same as measuring any other attribute. Iterating (or repeating) unit angles fills the spread between an angle's rays, just as iterating unit lengths fills a given length. Students discuss the need for a standard unit of measure, and they are introduced to the degree. An angle that measures 1 degree is a very small angle, which, when iterated 360 times, forms a circle.

Symmetry

Students complete symmetric figures that are partially given and create their own symmetric figures.

Multistep Multiplication Number Stories

Students continue solving multistep number stories, with a focus on multidigit multiplication strategies. They use number models that include a letter for the unknown, and they consider the reasonableness of their answers.

Please keep this Family Letter for reference as your child works through Unit 5.

125

Vocabulary

Important terms in Unit 5:

arc A part of a circle centered on the vertex of an angle. An arc is sometimes used to indicate where to measure the angle.

decompose To break apart a number or shape into smaller numbers or shapes.

degree A unit of measure for angles based on dividing a circle into 360 equal arcs.

full-turn A 360° rotation.

half-turn A 180° rotation.

like denominator A denominator that is the same in two or more fractions. For example, the fractions $\frac{3}{8}$, $\frac{5}{8}$, and $\frac{6}{8}$ have a like denominator, which is 8.

mirror image A point, line, or figure that exactly matches another point, line, or figure when it is reflected or folded over a line of symmetry so that it comes to rest on top of the corresponding image.

Sometimes the line of reflection is called a mirror, or mirror line.

mixed number A number that is written using both a whole number and a fraction.

quarter-turn A 90° rotation.

reflex angle An angle measure that is between 180° and 360°.

rotation A change in the direction an object faces; a turn.

straight angle An angle that measures 180°.

three-quarter turn A 270° rotation.

unit fraction A fraction whose numerator is 1. For example: $\frac{1}{2}$, $\frac{1}{3}$, and $\frac{1}{12}$ are unit fractions.

whole The entire object, collection of objects, or quantity being considered in a problem situation; 100%.

Do-Anytime Activities

To work with your child on concepts taught in this unit, try these activities:

1. Have your child help you measure when you are cooking or baking, using fractional measurements like $2\frac{1}{2}$ cups of flour or $\frac{1}{4}$ teaspoon of salt. Ask your child how you would double the measurements to make two batches instead of one. See whether he or she can show you one or two ways to do this.

2. Work with your child to create a line plot showing the number of hours family members spend sleeping or engaged in some other routine activity. Ask questions about the line plot; for example: "How many people in the family sleep for $8\frac{1}{2}$ hours?"

3. At home or when you are out together, encourage your child to point out items he or she believes are symmetric. Ask how many lines of symmetry there are in each of these objects.

4. Have your child point out angles in your home. Ask whether the angles are obtuse, acute, or right angles.

Building Skills through Games

In this unit, your child will play the following games as a way to increase his or her understanding of adding and subtracting fractions and mixed numbers, as well as angles, symmetry, and multistep multiplication number stories. For detailed instructions, see the *Student Reference Book*.

Decimal Top-It See *Student Reference Book*, page 253. This game provides practice comparing, ordering, reading, and identifying the values of digits in decimal numbers.

***Fishing for Fractions* (Addition/Subtraction)** See *Student Reference Book*, page 260. In this game students practice adding together two like fractions or subtracting one like fraction from another.

Fraction/Decimal Concentration See *Student Reference Book*, page 262. This game helps students recognize when fractions and decimals are equivalent.

Fraction Match See *Student Reference Book*, page 263. This game develops skill in naming equivalent fractions.

Fraction Top-It See *Student Reference Book*, page 265. This game develops skill in comparing fractions.

127

As You Help Your Child with Homework

As your child brings assignments home, it may be helpful to review the instructions together, clarifying them as necessary. The answers listed below will guide you through some of the Home Links in Unit 5.

Home Link 5-1

1. Sample answers:
$\frac{1}{5} + \frac{1}{5} + \frac{1}{5} + \frac{1}{5} + \frac{1}{5} + \frac{1}{5} + \frac{1}{5} + \frac{1}{5} + \frac{1}{5} + \frac{1}{5} + \frac{1}{5}$;
$\frac{5}{5} + \frac{5}{5} + \frac{1}{5}$;
$\frac{5}{5} + \frac{6}{5}$;
$1 + 1 + \frac{1}{5}$; $\frac{2}{5} + \frac{3}{5} + \frac{6}{5}$

3. a. Sample answer: $\frac{4}{12} + \frac{4}{12} = \frac{8}{12}$;

b. Sample answer: $\frac{8}{12} = \frac{2}{12} + \frac{2}{12} + \frac{2}{12} + \frac{2}{12}$;

5. 3,227   7. 1,950

Home Link 5-2

1. a.   b.   c.

3. a.

b.

c.

d.

5. 4

Home Link 5-3

1. a.
| Whole |
| new trees |

b. Sample answer: $\frac{1}{10} + \frac{3}{10} + \frac{2}{10} = t$

c. Sample answer:

| O | P | P | P | W | W | | | | |

d. $\frac{6}{10}$ oak, willow, or pine

3. $\frac{3}{5}$ 5. $\frac{10}{6}$, or $1\frac{4}{6}$

7. 0.4   9. 0.6

Home Link 5-4

1. a.
| Whole |
| ball of yarn |

b. Sample answer: $6\frac{2}{3} + 2\frac{2}{3} = s$

c. Sample answer: I decomposed the mixed numbers. Then I combined the wholes and the fractions. $6 + \frac{2}{3} + 2 + \frac{2}{3} = 8 + \frac{4}{3} = 9\frac{1}{3}$

d. $9\frac{1}{3}$, or $\frac{28}{3}$, balls

3. $8\frac{3}{6}$, or $\frac{51}{6}$

5. $6\frac{2}{4}$, or $\frac{26}{4}$ 7. 5,022   9. 1,092

Home Link 5-5

1. 20 hundredths + 15 hundredths = 35 hundredths

3. $\frac{10}{100} + \frac{50}{100} = \frac{60}{100}$, or $\frac{1}{10} + \frac{5}{10} = \frac{6}{10}$

5. $1 + \frac{30}{100} + 5 + \frac{64}{100} = 6\frac{94}{100}$

7. $\frac{150}{100} + \frac{78}{100} = \frac{228}{100}$, or $2\frac{28}{100}$

9. $\frac{2}{4}, \frac{3}{6}, \frac{4}{8}$ 11. $\frac{2}{8}, \frac{3}{12}, \frac{4}{16}$

Home Link 5-6

1. Sample answer: Bill and Carl didn't get $\frac{1}{5}$. They each got $\frac{1}{8}$. I know this because the two triangles are in $\frac{1}{4}$ of the whole land, so each is half of a fourth, or $\frac{1}{8}$.

3. 15,732   5. 10,591

Home Link 5-7

1. a.
Whole
Elijah's allowance

    b. Sample answer: $\frac{4}{5} - \frac{3}{5} = a$

    c. Sample answer:

    d. $\frac{1}{5}$ of his allowance

3. $\frac{1}{2}$ 5. $\frac{4}{5}$
7. 2,243   9. 17,437

Home Link 5-8

1. a.
Whole
1 cup

    b. Sample answer: $3\frac{1}{3} - 1\frac{2}{3} = c$

c. Sample answer: I started with what we had and counted up to what we needed. $1\frac{2}{3} + \frac{1}{3} = 2$. $2 + 1\frac{1}{3} = 3\frac{1}{3}$. Then I added: $\frac{1}{3} + 1\frac{1}{3} = 1\frac{2}{3}$.

d. $1\frac{2}{3}$, or $\frac{5}{3}$, cups

3. 1   5. $2\frac{3}{5}$, or $\frac{13}{5}$
7. 540   9. 8,084

Home Link 5-9

1.

3. $3\frac{1}{2} - \frac{1}{2} = d$; 3 inches

5. $\frac{3}{4}, \frac{6}{8}, \frac{9}{12}$ 7. $\frac{1}{2}, \frac{5}{10}, \frac{4}{8}$

Home Link 5-10

1–6. E, F, and G are sample answers

7. 4,250   9. 2,388

Home Link 5-11

1. angle *A*   **3.** angle *E*   **5.** angle *A* or *B*

7. 146,388   **9.** 12,961

Home Link 5-12

1. a. Triangle **b.** 2 sides **c.** 2 angles **d.** No.

3. 9   **5.** 6   **7.** 9

Home Link 5-13

1. Sample answer: $3 * (8 * 2 * 42) = p$;
Sample answer: $3 * 800 = \$2,400$;
$2,016, or 2,016 dollars

3. 9   **5.** 6   **7.** 36

Decomposing Fractions

Family Note In class today your child learned to decompose fractions into smaller parts. For example, $\frac{5}{6}$ can be decomposed into $\frac{1}{6} + \frac{1}{6} + \frac{1}{6} + \frac{1}{6} + \frac{1}{6}, \frac{2}{6} + \frac{3}{6}, \frac{1}{6} + \frac{4}{6}$, and so on.

Complete the name-collection boxes using equations.

SRB
125-127

① $\frac{11}{5}$

② $1\frac{3}{8}$

③ Decompose $\frac{8}{12}$ in more than one way into a sum of fractions with the same denominator.

Record each decomposition with an equation and justify it by shading the circle.

a. Equation: _____

b. Equation: _____

Practice

④ $9 * 785 =$ _____

⑤ $461 * 7 =$ _____

⑥ $644 * 4 =$ _____

⑦ _____ $= 39 * 50$

What Is the Whole?

For Problems 1-3, use your Geometry Template or sketch the shapes.

SRB
125-126

① Suppose [ ] is $\frac{1}{4}$. Draw each of the following:

Example: $\frac{3}{4}$ **a.** 1   **b.** $1\frac{1}{2}$ **c.** 2

② Suppose ◇ is $\frac{2}{3}$. Draw each of the following:

a. $\frac{1}{3}$ **b.** 1   **c.** $\frac{4}{3}$ **d.** 2

③ Suppose △ is $\frac{1}{3}$. Draw each of the following:

a. $\frac{3}{3}$ **b.** 2

c. $\frac{5}{3}$ **d.** $1\frac{1}{3}$

Practice

④ $\frac{4}{5} = \frac{8}{\boxed{}}$ ⑤ $\frac{3}{\boxed{}} = \frac{9}{12}$ ⑥ $\frac{9}{10} = \frac{\boxed{}}{100}$

133

Adding Fractions

Solve the number stories. Use a different strategy for each one.

SRB
47, 160–161

① The park department wants to have new trees planted. They agreed that $\frac{1}{10}$ of the trees will be oak, $\frac{3}{10}$ will be pine, and $\frac{2}{10}$ will be willow. They are undecided about the rest. What fraction of the trees will be oak, willow, or pine?

   a. Fill in the whole box.

        Whole

   b. Number model with unknown:

   c. One way to solve a fraction addition problem:

   d. Answer (with unit): _____

② The Patels have a DVD collection. Three-eighths of the DVDs are animated. Two-eighths of them are mysteries. One-eighth are comedies. The rest are about travel. What fraction of the DVDs are *not* about travel?

   a. Fill in the whole box.

        Whole

   b. Number model with unknown:

   c. A different way to solve a fraction addition problem:

   d. Answer (with unit): _____

Add.

③ $\frac{2}{5} + \frac{1}{5} = $ _____

④ $\frac{1}{2} + \frac{3}{2} = $ _____

⑤ $\frac{5}{6} + \frac{5}{6} = $ _____

⑥ $\frac{1}{3} + \frac{2}{3} + \frac{1}{3} = $ _____

Practice

Represent the fractions as decimals.

⑦ $\frac{4}{10} = $ _____

⑧ $\frac{40}{100} = $ _____

⑨ $\frac{6}{10} = $ _____

⑩ $\frac{6}{100} = $ _____

135

Mixed-Number Addition

Solve the number stories. Use a different strategy for each one.

SRB
47, 162-
163

① The art class had a box filled with balls of yarn. The students used $6\frac{2}{3}$ balls for a project. There are now $2\frac{2}{3}$ balls left in the box. How many balls of yarn did the art class start with?

a. Fill in the whole box.

Whole

b. Number model with unknown:

_____

c. One way to solve a mixed-number addition problem:

d. Answer (with unit): _____

② Mrs. Meyers is growing vines along the sides of her house. On the west side the vines are $2\frac{4}{10}$ meters tall. On the east side the vines are $5\frac{8}{10}$ meters taller than the ones on the west side. How tall are the vines on the east side?

a. Fill in the whole box.

Whole

b. Number model with unknown:

_____

c. A different way to solve a mixed-number addition problem:

d. Answer (with unit): _____

Add. Show your work.

③ $5\frac{2}{6} + 3\frac{1}{6} =$ _____

④ $1\frac{5}{8} + 2\frac{3}{8} =$ _____

⑤ $3\frac{3}{4} + 2\frac{3}{4} =$ _____

⑥ $3\frac{2}{5} + 1\frac{4}{5} + 2\frac{3}{5} =$ _____

Practice

⑦ $837 * 6 =$ _____

⑧ _____ $= 468 * 5$

⑨ _____ $= 364 * 3$

⑩ $56 * 70 =$ _____

137

Adding Tenths and Hundredths

Use what you know about equivalent fractions to add. Write an equation to show your work.

SRB
166-168

(1) 2 tenths + 15 hundredths

Equation (in words): _____

(2) $\frac{68}{100} + \frac{3}{10}$

Equation: _____

(3) $\frac{1}{10} + \frac{50}{100}$

Equation: _____

(4) $\frac{4}{10} + \frac{60}{100} + \frac{3}{10} + \frac{81}{100}$

Equation: _____

(5) $1\frac{3}{10} + 5\frac{64}{100}$

Equation: _____

(6) $3\frac{22}{100} + 2\frac{8}{10}$

Equation: _____

(7) $\frac{15}{10} + \frac{78}{100}$

Equation: _____

(8) Nicholas shaded $\frac{40}{100}$ of his hundreds grid. Victor shaded $\frac{5}{10}$ of his grid.

Who shaded more? _____

How much did they shade in all? _____ of a grid

Practice

Write three equivalent fractions.

(9) $\frac{1}{2}$ = _____

(10) $\frac{1}{3}$ = _____

(11) $\frac{1}{4}$ = _____

(12) $\frac{1}{5}$ = _____

Fraction Error Finder

Consider this problem:

A king owns land outside of his castle.

He has partitioned the land to give as gifts to his 5 sons.

What fraction of the land did the king give to each of his sons?

Here is Zeke's solution:

Andy got $\frac{1}{2}$ Bill got $\frac{1}{5}$ Carl got $\frac{1}{5}$

Dirk got $\frac{1}{8}$ Evan got $\frac{1}{8}$

① Identify Zeke's two errors, correct them, and explain why your answer is correct.

② Write a fraction addition equation to represent the correct answers and show the sum of the pieces of land.

_____

Practice

Use U.S. traditional addition and subtraction.

③ 8,936 + 6,796 = _____   ④ 635 − 392 = _____

⑤ 6,386 + 4,205 = _____   ⑥ 900 − 463 = _____

141

Subtracting Fractions

Solve the number stories. Use a different strategy for each one.

SRB
160-161

① Elijah still had $\frac{4}{5}$ of his allowance at the end of the month. Then he spent $\frac{3}{5}$ of his original allowance on a movie ticket and popcorn. How much of Elijah's allowance was left?

a. Fill in the whole box.

Whole

b. Number model with unknown: _____

c. One way to solve a fraction subtraction problem:

d. Answer (with unit): _____

② Kendra's computer battery had $\frac{9}{10}$ of a charge. After her sister Lydia borrowed the computer, the battery had $\frac{3}{10}$ of a charge left. How much of the battery charge did Lydia use?

a. Fill in the whole box.

Whole

b. Number model with unknown: _____

c. Another way to solve a fraction subtraction problem.

d. Answer (with unit): _____

Subtract.

③ $\frac{2}{2} - \frac{1}{2} = $ _____

④ $\frac{11}{6} - \frac{4}{6} = $ _____

⑤ _____ $= 1 - \frac{1}{5}$

Practice

⑥ $8,936 + 6,796 = $ _____

⑦ _____ $= 4,635 - 2,392$

⑧ _____ $= 46,386 + 4,205$

⑨ $65,900 - 48,463 = $ _____

143

Mixed-Number Subtraction

Solve the number stories. Use a different strategy for each one.

SRB
164-165

(1) The chocolate chip cake recipe calls for $3\frac{1}{3}$ cups of milk. We only have $1\frac{2}{3}$ cups at home. How much more milk do we need?

Whole

a. Fill in the whole box.

b. Number model with unknown: _____

c. One way to solve a mixed-number subtraction problem:

d. Answer (with unit): _____

(2) Lourdes is listening to an audio book that is 9 hours long. She has listened for $6\frac{1}{6}$ hours so far. How many hours of listening time are left?

Whole

a. Fill in the whole box.

b. Number model with unknown: _____

c. A different way to solve a mixed-number subtraction problem:

d. Answer (with unit): _____

Subtract. Show your work.

(3) $4\frac{1}{2} - 3\frac{1}{2} =$ _____

(4) _____ $= 5\frac{8}{12} - 5\frac{3}{12}$

(5) $4\frac{2}{5} - 1\frac{4}{5} =$ _____

(6) _____ $= 9\frac{4}{10} - 3\frac{8}{10}$

Practice

(7) _____ $= 54 * 10$

(8) $63 * 100 =$ _____

(9) $86 * 94 =$ _____

(10) $5,715 * 6 =$ _____

145

Student Growth

Mrs. Welch surveyed her students about how much they had grown over the past year. This is the data she gathered.

Student Growth Over the Past Year (to the nearest $\frac{1}{2}$ inch)	
$1\frac{1}{2}$	$1\frac{1}{2}$
2	$2\frac{1}{2}$
$2\frac{1}{2}$	2
$\frac{1}{2}$	$1\frac{1}{2}$
$2\frac{1}{2}$	$\frac{1}{2}$
1	2
$1\frac{1}{2}$	2
$1\frac{1}{2}$	$\frac{1}{2}$
$3\frac{1}{2}$	$1\frac{1}{2}$
1	1
1	$2\frac{1}{2}$
2	2
$2\frac{1}{2}$	$1\frac{1}{2}$

(1) Plot the data set on the line plot.

_____
Title

<————+———+———+———+———+———+———+———+————>

___ ___ ___ ___ ___ ___ ___ ___

_____
Label

Use the completed line plot to answer the questions.

(2) What is the greatest number of inches a student grew in a year?

About _____ inch(es)   The least? About _____ inch(es)

(3) What is the difference between the greatest and the least number of inches grown?

Number model with unknown: _____   Answer: _____ inch(es)

Practice

Circle the three equivalent fractions in each group.

(4) $\frac{1}{4}$, $\frac{3}{6}$, $\frac{1}{8}$, $\frac{2}{8}$, $\frac{3}{12}$

(5) $\frac{3}{4}$, $\frac{4}{8}$, $\frac{6}{8}$, $\frac{5}{6}$, $\frac{9}{12}$

(6) $\frac{2}{3}$, $\frac{1}{5}$, $\frac{4}{6}$, $\frac{7}{12}$, $\frac{8}{12}$

(7) $\frac{1}{2}$, $\frac{5}{10}$, $\frac{4}{8}$, $\frac{7}{12}$

Rotations

Family Note If your child needs help with the following problems, consider putting up signs in a room in your home to indicate the directions *north, south, east,* and *west.* Do the turns with your child.

Please return this Home Link to school tomorrow.

left turn
counterclockwise   right turn
clockwise

Make the turns described below. Show which way you face after each turn by:

- Drawing a dot on the circle.

- Labeling the dot with a letter.

Example: Face north.

Do a $\frac{1}{2}$ turn counterclockwise.

On the circle, mark the direction you are facing with the letter *A.*

SRB
207

North

West   •   East
You

A
South

① Face north. Do a $\frac{1}{4}$ turn clockwise. Mark the direction you are facing with the letter *B.*

② Face north. Do a $\frac{3}{4}$ turn clockwise. Mark the direction you are facing with the letter *C.*

③ Face east. Do a $\frac{1}{4}$ turn counterclockwise. Mark the direction you are facing with the letter *D.*

④ Face west. Make less than a $\frac{1}{4}$ turn clockwise. Mark the direction you are facing with the letter *E.*

⑤ Face north. Make a clockwise turn that is more than a $\frac{1}{2}$ turn but less than a $\frac{3}{4}$ turn. Mark the direction you are facing with the letter *F.*

⑥ Face north. Make a counterclockwise turn that is less than a $\frac{1}{2}$ turn but more than a $\frac{1}{4}$ turn. Mark the direction you are facing with the letter *G.*

Practice

⑦ 85 * 50 = _____

⑧ 416 * 6 = _____

⑨ _____ = 597 * 4

⑩ 1,373 * 7 = _____

Estimating Angle Measures

Family Note Our class has been learning about turns, angles, and angle measures. A full turn can be represented by an angle of 360°, a $\frac{1}{2}$ turn by an angle of 180°, a $\frac{1}{4}$ turn by an angle of 90°, and so on. Help your child match the measures below with the angles pictured. (It is not necessary to measure the angles with a protractor.)

Name which angle has the given measure.

(1) about 180°   angle _____

(2) about 90°   angle _____

(3) about 270°   angle _____

(4) between 0° and 90°   angle _____

(5) between 90° and 180° angle _____

Rotation	Degrees
$\frac{1}{4}$ turn	90°
$\frac{1}{2}$ turn	180°
$\frac{3}{4}$ turn	270°
full turn	360°

SRB
207

Practice

(6) 5,956 + 4,983 = _____

(7) 60,351 + 86,037 = _____

(8) 41,015 − 517 = _____

(9) 23,730 − 10,769 = _____

Folding Shapes

Family Note Our class has been studying lines of symmetry—lines that divide figures into mirror images. Help your child look for symmetric shapes in books, newspapers, and magazines, and in objects around the house, such as windows, furniture, dishes, and so on.

Please bring your cutouts to school tomorrow.

① Fold a sheet of paper in half. Cut off the folded corner, as shown. Before you unfold the cutoff piece, guess its shape.

SRB
238

     a. Unfold the cutoff piece. What shape is it?

     _____

     b. How many sides of the cutoff piece are the same length? _____ sides

     c. How many angles are the same size? _____ angles

     d. The fold is a line of symmetry. Does the cutoff piece have any other lines of symmetry? _____

② Fold another sheet of paper in half. Fold it in half again. Make a mark on both folded edges 2 inches from the folded corner. Cut off the folded corner. Before you unfold the cutoff piece, guess its shape.

2 in. {

2 in.

     a. Unfold the cutoff piece. What shape is it? _____

     b. Are there any other lines of symmetry besides the fold lines? _____

     c. On the back of this paper, draw a picture of the cutoff shape. Draw all of its lines of symmetry.

Practice

③ $81 \div$ _____ $= 9$ ④ _____ $\div 9 = 6$

⑤ $7 = 42 \div$ _____   ⑥ _____ $\div 9 = 8$

⑦ $36 \div$ _____ $= 4$ ⑧ $8 =$ _____ $\div 6$

Expressing Answers to Number Stories

Family Note Today students learned to express solutions to multistep number stories using correct units and single number models. Have your child explain the steps for solving each of the problems below, and then help him or her write these steps as a single number model, including a letter for the unknown quantity.

Solve. Record a long number model with a letter for the unknown quantity and write the answer with the correct unit.

① Guillermo hires two painters to paint the walls of his living room. The painters each make $42 an hour for an 8-hour workday. If the work takes 3 days, how much will Guillermo pay the painters?

Number model with unknown: _____

Estimate: _____

Answer (with unit): _____

② Blaine is on vacation in New York City and wants to collect magnets of places he visits to give to all his friends. The Times Square magnets cost $2 each and come in sets of 4. The Statue of Liberty magnets cost $3 each and come in sets of 5. If Blaine buys 12 sets of each type of magnet, how much will he pay?

Number model with unknown: _____

Estimate: _____

Answer (with unit): _____

Practice

③ 45 ÷ 5 = _____   ④ 56 ÷ 8 = _____   ⑤ 54 ÷ 9 = _____

⑥ _____ ÷ 9 = 4   ⑦ _____ ÷ 6 = 6   ⑧ _____ ÷ 8 = 3

155

Division; Angles

Division

In Unit 6 your child will divide multidigit numbers using extended division facts, multiples, area models, and partial quotients. Working with more than one division strategy helps students build conceptual knowledge and means that they have more than just one method to choose from. Throughout the unit students solve multistep division number stories involving dividends with multiple digits, learn the meaning of the remainders, and apply their division skills in real-life contexts.

The unit begins with extended division facts. Knowing that $24 \div 4 = 6$ enables students to see that $240 \div 4 = 60$; $240 \div 40 = 6$; $2,400 \div 4 = 600$; and so forth. Students play *Divide and Conquer*, where they practice dividing with extended facts. The confidence they build by working with extended division facts will help them to divide larger numbers with ease.

Students also learn the partial-quotients division method, in which the dividend is divided in a series of steps. The first example below illustrates a model of the partial-quotients method for $1,325 \div 9$. When students partition, or divide, the 1,325 into parts ($900 + 360 + 63 + 2$), it helps them develop their understanding of the algorithm. The second example uses the partial-quotients method. The quotients for each step are added together to give the final answer.

	9
1,325	*s*

$100 * 9 = 900$	100
$40 * 9 = 360$	40
$7 * 9 = 63$	7

```
       1325
     −  900
 +
        425
     −  360
 +
         65
     −   63
 ──────────
 147      2
```

```
  9)1,325
   −  900    100
      425
   −  360     40
       65
   −   63      7
        2    147
```

Angles

Students continue their work with angle measurement and learn to use both full-circle and half-circle protractors. They learn that angle measurements can be added, and they use this understanding and properties of angles to find unknown angle measures.

Fraction Operations

Students continue working with addition and subtraction of fractions and mixed numbers. They apply their knowledge of multiplication to explore multiplying a fraction by a whole number.

Please keep this Family Letter for reference as your child works through Unit 6.

Vocabulary

Important terms in Unit 6:

complementary angles Angles with measures that equal 90° when added together.

extended division facts Variations of division facts involving multiples of 10, 100, and so on. For example, 720 ÷ 8 = 90 is an extended fact related to 72 ÷ 8 = 9.

partial quotients A way to divide in which the dividend is divided in a series of steps. The quotients for each step (called partial quotients) are added to give the final answer.

protractor A tool that measures angles in degrees.

reflex angle An angle measure that is between 180° and 360°.

straight angle An angle that measures 180°.

supplementary angles Angles with measures that equal 180° when added together.

Do-Anytime Activities

To work with your child on concepts taught in this unit, try these activities.

1. Practice extended division facts, such as 1,800 ÷ 30.

2. Ask your child to help you divide something for dinner into equal portions for each member of your family. For example, ask, "How can we divide the 5 chicken breasts equally for the 4 of us?"

3. Ask questions like these:
   - What kind of angles do you see on a stop sign?
   - What types of angles are on our tile or wood floors, or on the walls?
   - What types of angles are in a rectangular sign?
   - What types of angles do you see in the supports for the bridge?

4. Make up some situations such as those listed directly above, and encourage your child to draw a picture or diagram to show you how to solve it.

Building Skills through Games

In this unit your child will play the following games to increase his or her understanding of division and angles. For detailed instructions on how to play these games, please see the *Student Reference Book*.

Angle Add-Up See *Student Reference Book*, page 248. This game provides practice adding and subtracting angle measures.

Divide and Conquer See *Student Reference Book*, page 254. This game for three players—the Caller, the Brain, and the Calculator—provides practice with extended division facts.

As You Help Your Child with Homework

As your child brings assignments home, it may be helpful to review the instructions together, clarifying them as necessary. The answers listed below will guide you through some of the Home Links in Unit 6.

Home Link 6-1

1. 4; 40

3. **a.** 5 **b.** 50 **c.** 500 **d.** 5

5. **a.** 2 **b.** 20 **c.** 200 **d.** 2

7. 2,280 9. 6,335

Home Link 6-2

1. Sample answer: $2 * s = 60$; 30 meters

3. 3; 7; 45; 10 5. 60 7. 60

Home Link 6-3

1. 40, 42, 44, 46, 48, 50;
   $46 / 2 = b$; 23 packages; $46 / 2 = 23$

3. 820 5. 999

Home Link 6-4

1. Sample estimate: $45 / 3 = 15$; $48 \div 3 = p$;
   16 pounds

3. Sample answer: $\frac{3}{6}$, $\frac{4}{8}$

5. Sample answer: $\frac{1}{4}$, $\frac{3}{12}$

Home Link 6-5

1. Sample answer: 115 is the total number of students. 4 is the number of buses. 28 is the number of students per bus. 3 is the number of students left over after dividing evenly.

2. Sample answer: Because 28 students from each class can be on a bus and there are 3 students left over, 3 buses will have 29 students. Then, because each bus needs a teacher, 3 buses will have 30 passengers on them and 1 bus will have 29 passengers.

Mr. Atkins's class has too many students to fit on one bus. So he can go on the bus with most of his students, and 2 students will have to ride on another bus. His bus will have 30 passengers.

Mrs. Gonzales's class has the fewest students. Because she has 27 students and adding herself makes 28 passengers, her bus will have room for Mr. Atkins's 2 extra students.

Mr. Bates and his students are a perfect fit for a bus. There will be 30 passengers on his bus.

Ms. Smith and her students fit on a bus, with room for one more. However, that spot is not needed.

3. $\frac{7}{8}$ 5. $\frac{2}{5}$

Home Link 6-6

1. 12,000; 7; 16,000, 11 3. 8,000 pounds

5. $\frac{7}{8}$ 7. $\frac{53}{100}$

Home Link 6-7

1. Sample answer:

$$
\begin{array}{r|l}
5)\overline{360} & \\
-350 & 70 \\
\hline
10 & \\
-10 & 2 \\
\hline
0 & 72 \\
\end{array}
$$

Sample estimate: $350 \div 5 = 70$; $360 \div 5 = p$; 72 prizes; 0 prizes

3. Sample estimate: $160 / 8 = 20$; 23

5. 0.08, 0.34, 0.98, 9.8 7. >

159

Home Link 6-8

1.

Sample number models are given.

$124 / 8 = s$; $15\frac{4}{8}$, or $15\frac{1}{2}$ strawberries;
$124 / 8 \rightarrow 15$ R4;

B. Reported it as a fraction;
Sample answer: You can cut the remaining strawberries into halves.

3. $\frac{3}{8}, \frac{3}{6}, \frac{3}{5}, \frac{3}{3}$ 5. $\frac{1}{2}, \frac{2}{3}, \frac{6}{8}, \frac{99}{100}$

Home Link 6-9

1. Right; 90°   3. Acute; 45°

5. 692   7. 680

Home Link 6-10

1. 60°   3. 84°

5.

7. 65,811   9. 64,091

Home Link 6-11

1. Sample answer: $30° + y = 90°$; 60°

3. Sample answer: $90° - z = 75°$; 15°

5. Sample answer: $180° - 60° = a$; 120°

7. $\frac{7}{12}, \frac{7}{10}, \frac{7}{9}, \frac{7}{8}$

Home Link 6-12

1. **a.** Strawberries; $\frac{3}{12} + \frac{1}{12} = b$; $\frac{4}{12}$ pound
   b. $\frac{3}{12} - \frac{1}{12} = p$; $\frac{2}{12}$ pound

3. $4\frac{2}{8} + 1\frac{3}{8} = p$; $5\frac{5}{8}$ pounds

5. 2,400

Home Link 6-13

1. 45 children; Sample answer:

5 groups of 9;
$9 + 9 + 9 + 9 + 9 = 45$; $5 * 9 = 45$

3. $2\frac{2}{5}$ veggie pizzas; Sample answer:

4 groups of $\frac{3}{5}$;
$\frac{3}{5} + \frac{3}{5} + \frac{3}{5} + \frac{3}{5} = \frac{12}{5}$; $4 * \frac{3}{5} = \frac{12}{5}$

5. 19

Solving Extended Division Facts

Write a basic division fact and an extended division fact for each Fact Triangle.

SRB 110

①

160

×, ÷

4

Basic fact: $16 \div 4 =$ _____

Extended fact: $160 \div 4 =$ _____

②

240

×, ÷

4

Basic fact: _____

Extended fact: _____

Solve.

③ **a.** $25 \div 5$ _____

b. $250 \div 5 =$ _____

c. $2,500 \div 5 =$ _____

d. $250 \div 50 =$ _____

④ **a.** $36 \div 4 =$ _____

b. $360 \div 4 =$ _____

c. $3,600 \div 4 =$ _____

d. $360 \div 40 =$ _____

⑤ **a.** $18 \div 9 =$ _____

b. $180 \div 9 =$ _____

c. $1,800 \div 9 =$ _____

d. $180 \div 90 =$ _____

⑥ **a.** $42 / 7 =$ _____

b. $420 / 7 =$ _____

c. $4,200 / 7 =$ _____

d. $420 / 70 =$ _____

Practice

⑦ $456 * 5 =$ _____

⑧ $720 * 8 =$ _____

⑨ $905 * 7 =$ _____

Finding the Unknown Side Length

Solve.

(1)

s

2 meters | 60 square meters

How long is the unknown side s?

Equation with unknown: _____

Answer: _____ meters

(2)

t

6 meters | 420 square meters

What is the length of the unknown side t?

Equation with unknown: _____

Answer: _____ meters

(3) Fill in the unknown information about some rectangular rooms in a museum.

Room	Length in Yards	Width in Yards	Area in Square Yards
A	6		18
B		8	56
C	9	5	
D		9	90

(4) A store is rectangular in shape with an area of 2,700 square feet. It has a length of 90 feet. How wide is it?

Equation with unknown: _____

Answer: _____ feet

Practice

(5) $420 \div 7 =$ _____

(6) _____ $= 3,600 / 6$

(7) $5,400 \div 90 =$ _____

Solving Division
Number Stories

Fill in the lists of multiples to help you, if needed.

SRB
55, 110,
114

① Rosario sells bicycle wheels in packages of 2. If a store orders 46 wheels, how many packages will she send?

20 [2s] = _____   Number model with unknown: _____

21 [2s] = _____   Answer: _____ packages

22 [2s] = _____   Number model with answer: _____

23 [2s] = _____

24 [2s] = _____

25 [2s] = _____

② Doug is placing apples in bags for a picnic. He can fit 6 apples in a bag. If he has 92 apples, how many bags will he need?

10 [6s] = _____   Number model with unknown: _____

11 [6s] = _____   Answer: _____ bags

12 [6s] = _____   Number model with answer: _____

13 [6s] = _____

14 [6s] = _____

15 [6s] = _____

16 [6s] = _____

17 [6s] = _____

18 [6s] = _____

Practice

③ $82 * 10 =$ _____   ④ _____ $= 25 * 30$ ⑤ $333 * 3 =$ _____

Partial-Quotients Division

Family Note In this lesson students are introduced to the partial-quotients method, in which a number is divided in a series of steps. The quotients for each step (called partial quotients) are added to give the final answer. For example, to divide 96 by 6, students use extended multiplication facts such as 6 * **10** = 60 to find the partial quotient. Then with the remaining 36, they use an "easy" multiplication fact, 6 * **6**, to finish solving the problem. These two partial quotients are added together, 10 + 6, to find the exact quotient of 16. So 96 ÷ 6 = 16.

Estimate. Write a number model with an unknown to represent the problem. Then solve using partial quotients.

SRB
47,
111-114

(1) Jordan has 3 Great Dane puppies. At 6 weeks old, their combined weight is 48 pounds. Assuming that they all weigh about the same amount, how much does each puppy weigh?

Estimate: _____

Number model with unknown: _____

Answer: _____ pound(s)

(2) Four sisters love barrettes. They have a value pack that contains 92 barrettes. How many barrettes can each sister have if they share equally?

Estimate: _____

Number model with unknown _____

Answer: _____ barrette(s)

Practice

Name two equivalent fractions for each fraction given.

(3) $\frac{1}{2}$ ____ ____   (4) $\frac{1}{3}$ ____ ____   (5) $\frac{25}{100}$ ____ ____   (6) $\frac{6}{8}$ ____ ____

Assigning People to Buses

Mr. Atkins is organizing the 4th- and 5th-grade field trip to the science museum. He asked his students to help him figure out which students and teachers should go on each bus. The number of students in each class is shown in the table below:

SRB
113-115

Mr. Atkins's 4th-grade class	31 students
Ms. Smith's 4th-grade class	28 students
Mr. Bates's 5th-grade class	29 students
Mrs. Gonzales's 5th-grade class	27 students

Important information:

- 4 buses have been ordered.
- The maximum number of passengers is 30 per bus.
- Each bus must have 1 teacher.

Cary said he solved the problem this way:

115 / 4 is 28 with a remainder of 3.

① What do the numbers in his sentence mean?

② Which students and teachers should go on each bus? Explain why.

Practice

③ $\frac{3}{8} + \frac{4}{8} =$ _____ ④ $\frac{5}{6} + \frac{3}{6} =$ _____ ⑤ $\frac{4}{5} - \frac{2}{5} =$ _____ ⑥ $\frac{7}{10} - \frac{3}{10} =$ _____

Converting Units of Weight

tons 0 ——— 1 ——— 2 ——— 3 ——— 4 ——— 5

pounds 0 — 2,000 — 4,000 — 6,000 — 8,000 — 10,000

SRB
190-192

pounds 0 — 1 — 2 — 3 — 4 — 5 — 6 — 7 — 8 — 9 — 10

ounces 0 — 16 — 32 — 48 — 64 — 80 — 96 — 112 — 128 — 144 — 160

Use the measurement scales to help you solve the problems.

①

Tons	Pounds
1	2,000
6	
	14,000
8	
	22,000

②

Pounds	Ounces
1	16
5	
9	
	160
15	

③ The army chef is ordering food for the troops. She ordered 2 tons of rice, 1 ton of pasta, and 1 ton of flour. How many pounds of food did she order?

Answer: _____ pound(s)

④ Potatoes come in 8-pound bags. How many ounces do 12 bags weigh?

Answer: _____ ounce(s)

Practice

⑤ $\frac{4}{8} + \frac{3}{8} =$ _____   ⑥ _____ $= \frac{5}{8} - \frac{3}{8}$ ⑦ _____ $= \frac{5}{10} + \frac{3}{100}$ ⑧ $\frac{60}{100} + \frac{4}{10} =$ _____

171

Partial Quotients

Estimate. Write a number model to represent the problem. Solve using partial quotients. SRB
47,
113-114

(1) The carnival committee has 360 small prizes to distribute to 5 booths. How many prizes will each booth get?

Estimate: _____

Number model with unknown:

_____

(2) The mall needs a row of parking spaces. The length of the parking area is 2,711 feet. If each parking space is 9 feet wide, how many spaces will there be?

Estimate: _____

Number model with unknown:

_____

Answer: _____ prizes

How many prizes are left over? ___ prizes

Answer: _____ spaces

How many feet are left over? ___ feet

Solve using partial quotients. Show your work on the back of this page.

(3) 161 / 7 Estimate: _____   Answer: _____

(4) 576 / 4 Estimate: _____   Answer: _____

Practice

Put these decimals in order from least to greatest.

(5) 0.98, 0.34, 9.8, 0.08 _____, _____, _____, _____

(6) 0.11, 0.01, 0.10, 1.0 _____, _____, _____, _____

Use <, >, or = to compare the decimals.

(7) 0.65 _____ 0.5

(8) 37.9 _____ 37.96

173

Interpreting Remainders

① Mrs. Patel brought a box of 124 strawberries to the party. She wants to divide the strawberries evenly among 8 people. How many strawberries will each person get?

② Mr. Chew has a box of 250 pens. He asks Maurice to divide the pens into groups of 8. How many groups can Maurice make?

Number model with unknown:

_____

Answer:

_____ strawberries

Number model with answer:

_____

What did you do about the remainder? Circle the answer.

A. Ignored it

B. Reported it as a fraction

C. Rounded the answer up

Why? _____

_____

_____

Number model with unknown:

_____

Answer:

_____ groups

Number model with answer:

_____

What did you do about the remainder? Circle the answer.

A. Ignored it

B. Reported it as a fraction

C. Rounded the answer up

Why? _____

_____

_____

Practice

Order the fractions from smallest to largest.

③ $\frac{3}{6}, \frac{3}{3}, \frac{3}{5}, \frac{3}{8}$ ____, ____, ____, ____

④ $\frac{1}{4}, \frac{1}{8}, \frac{1}{2}, \frac{1}{5}$ ____, ____, ____, ____

⑤ $\frac{2}{3}, \frac{1}{2}, \frac{6}{8}, \frac{99}{100}$ ____, ____, ____, ____

⑥ $\frac{4}{5}, \frac{81}{100}, \frac{4}{6}, \frac{2}{10}$ ____, ____, ____, ____

Measuring Angles

Cut out the angle measurer and use a pencil to poke a hole through the center.

Label each angle *acute*, *right*, or *obtuse*.

Then use the angle measurer to measure each angle.

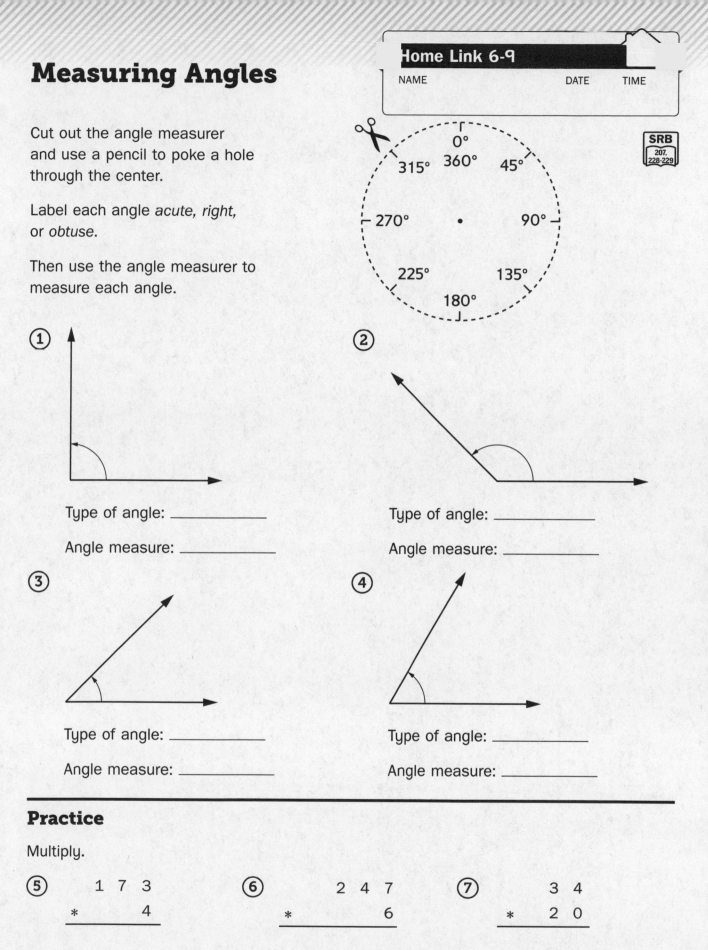

SRB
207,
228-229

① Type of angle: _____

Angle measure: _____

② Type of angle: _____

Angle measure: _____

③ Type of angle: _____

Angle measure: _____

④ Type of angle: _____

Angle measure: _____

Practice

Multiply.

⑤   1 7 3
    *   4
_____

⑥   2 4 7
    *   6
_____

⑦   3 4
    *   2 0
_____

177

Measuring Angles with a Protractor

First estimate whether the angles measure more or less than 90°. Then use a half-circle protractor to measure them.

① ∠A: _____°   ② ∠B: _____°   ③ ∠C: _____°

④ ∠QRS: 105°   ⑤ ∠NOP: 32°   ⑥ ∠KLM: 300°

Practice

⑦   2 3, 8 0 7
   + 4 2, 0 0 4
   _____

⑧   5 3 0, 0 8 3
   + 2 8 3, 6 9 0
   _____

⑨   8 7, 9 4 2
   − 2 3, 8 5 1
   _____

⑩   6 0 0, 2 9 9
   − 5 1 0, 3 4 5
   _____

179

Finding Angle Measures

Find the unknown angle measures in Problems 1–6. Do *not* use a protractor.

A **right angle** measures 90°.

A **straight angle** measures 180°.

(1)

30°

y

Equation with unknown: _____

Answer: _____ °

(2)

x

55°

Equation with unknown: _____

Answer: _____ °

(3)

75°

z

Equation with unknown: _____

Answer: _____ °

(4)

b   100°

Equation with unknown: _____

Answer: _____ °

(5)

a   60°

Equation with unknown: _____

Answer: _____ °

(6)

p

70°

Equation with unknown: _____

Answer: _____ °

Practice

Order the fractions from smallest to largest.

(7) $\frac{7}{10}, \frac{7}{8}, \frac{7}{12}, \frac{7}{9}$ _____

(8) $\frac{5}{9}, \frac{99}{100}, \frac{1}{4}, \frac{9}{10}$ _____

181

Solving Number Stories

Write a number model with an unknown to represent each problem. Then solve.

SRB
47,
162-165

1. Martin had some leftover fruit from making fruit salad. He had $\frac{3}{12}$ pound of strawberries and $\frac{1}{12}$ pound of blueberries.

   Which fruit weighed more? _____

   a. How many pounds of fruit did Martin have left?

      Number model with unknown: _____

      Answer: _____ pound

   b. How much more did the strawberries weigh than the blueberries?

      Number model with unknown: _____

      Answer: _____ pound

2. Charlotte and Beth each made a vegetable salad to take to a reunion. Together the salads weighed 6 pounds. Charlotte's salad weighed $3\frac{1}{2}$ pounds.

   a. How much did Beth's salad weigh?

      Number model with unknown: _____

      Answer: _____ pounds

   b. How much more did Charlotte's salad weigh than Beth's?

      Number model with unknown: _____

      Answer: _____ pound

3. Andy's potato salad weighed $1\frac{3}{8}$ pounds more than Mardi's. Mardi's potato salad weighed $4\frac{2}{8}$ pounds. How much did Andy's potato salad weigh?

   Number model with unknown: _____

   Answer: _____ pounds

Practice

4. $826 * 5 =$ _____

5. $48 * 50 =$ _____

Multiplying a Fraction by a Whole Number

SRB
173-174

Solve. Use drawings, words, and equations to represent the problems.

① 5 vans were needed for a camp field trip. There were 9 children per van.

How many children went on the field trip? _____ children

Drawing:

Words: _____ groups of _____

Addition equation: _____

Multiplication equation: _____

② Penny and her 2 friends each ate $\frac{1}{6}$ of a cake. How much cake did they eat?

_____ of a cake

Drawing:

Words: _____ groups of _____

Addition equation: _____

Multiplication equation: _____

③ Christopher wants to give his 4 friends $\frac{3}{5}$ of a veggie pizza each.

How much veggie pizza will he need? _____ veggie pizzas

Drawing:

Words: _____ groups of _____

Addition equation: _____

Multiplication equation: _____

Practice

④ $84 / 6 =$ _____

⑤ $76 \div 4 =$ _____

⑥ _____ $= 90 \div 5$

185

Multiplication of a Fraction by a Whole Number; Measurement

Fractions

Unit 7 begins with students applying and extending their previous understandings of multiplying whole numbers to multiplying a fraction by a whole number. Your child will multiply fractions by whole numbers in different ways: using concrete objects, drawing pictures, and writing equations. Using a variety of strategies helps students build conceptual knowledge and gives them more than one method to choose from when solving problems.

Consider this number story, for example: *Mattie needs $\frac{1}{2}$ cup of granola for each member of her family. She has 5 family members. How much granola does she need for everyone in the family?*

Below are examples of different strategies students might use to solve the problem.

- Use repeated addition: $\frac{1}{2} + \frac{1}{2} + \frac{1}{2} + \frac{1}{2} + \frac{1}{2} = \frac{5}{2}$, or $2\frac{1}{2}$ cups of granola

- Apply relational thinking: Two $\frac{1}{2}$ s are 1. Four $\frac{1}{2}$ s are two. Another $\frac{1}{2}$ is $2\frac{1}{2}$.

- Draw a picture:

- Use fraction circles:

- Use equations: $5 * \frac{1}{2} = \frac{(5 * 1)}{2} = \frac{5}{2}$

In this unit students create drawings or use models, such as fraction circles or fraction strips, to explain their thinking as they apply their skills in real-life contexts involving time, weight, capacity, and money.

Measurement

In Unit 7 students work with increasingly complex measurement conversion problems. They explore U.S. customary units of capacity, including the cup, pint, quart, and gallon, and solve number stories involving conversions between whole numbers and fractions of units.

Students also convert between pounds and ounces in the course of solving real-world number stories involving U.S. customary units of weight. Lesson 7-12 challenges students with number stories involving decimals in a money context. Here they apply their understanding of fraction/decimal equivalencies and fraction operations to solve the problems. According to the Common Core State Standards, students are not expected to perform operations with decimals until fifth grade. However, the link established through these activities between different representations of numbers, especially fractions and decimals, is a key prerequisite concept for success with decimal computation. Problems like the ones presented in this unit build the foundation for that later work.

Line Plots

Line plots are used to organize and display data. Students analyze data measured to an eighth of a unit, create their own line plots, and use line plots to solve problems involving computations with fractions and mixed numbers.

Division

Students estimate, solve, and assess the reasonableness of answers to multistep division number stories. They plan strategies and write number models with letters for the unknown quantities, explaining how they found each answer and checking to make sure their answer makes sense. Students use division strategies to solve real-world measurement number stories, converting between different units of measurement.

Please keep this Family Letter for reference as your child works through Unit 7.

Vocabulary

Important terms in Unit 7:

line plot A sketch of data in which checkmarks, Xs, stick-on notes, or other marks above a labeled line show the frequency of each value.

mixed number A number that is written using both a *whole number* and a *fraction*. For example, $5\frac{2}{3}$ is a mixed number equal to $5 + \frac{2}{3}$.

multiple of a fraction A product of a fraction and a counting number. For example, $\frac{5}{4}$ is a multiple of $\frac{1}{4}$ because $\frac{5}{4} = 5 * \left(\frac{1}{4}\right)$.

unit fraction A fraction in which the numerator is 1. For example, $\frac{1}{4}$, $\frac{1}{6}$, and $\frac{1}{10}$ are unit fractions. Fractions can be built from unit fractions. For example, $\frac{3}{4}$ can be built from three $\frac{1}{4}$s.

Do-Anytime Activities

To work with your child on concepts taught in this unit, try these activities:

1. Have your child make a list of shoe sizes from the members of the household and create a line plot from the data. Ask questions like these: *What is the largest shoe size? The smallest? What is the difference between the largest and smallest shoe size?*

2. Ask your child to convert weights of common items into fractions of a pound. For example, a 4-ounce tube of toothpaste $= \frac{1}{4}$ pound.

3. Ask questions like these:
   - *How long did it take you to get to school?*
   - *What fraction of an hour is that?*
   - *If it takes you 3 times as long to get to school tomorrow, how long will it take you?*
   - *How much time do you spend all week getting to school?*

4. Look at a store advertisement or sale flyer and pose questions about items sold in bulk. For example: *What is the cost of 1 _____? What is the cost if we buy _____ or _____?*

Building Skills through Games

In this unit your child will play the following new game to increase his or her understanding of fraction operations. For detailed instructions, see the *Student Reference Book.*

Fraction Multiplication Top-It See *Student Reference Book*, page 264. Students practice multiplying a whole number by a fraction, and they compare their answer with a partner's.

As You Help Your Child with Homework

As your child brings assignments home, it may be helpful to review the instructions together, clarifying them as necessary. The answers listed below will guide you through the Home Links in Unit 7.

Home Link 7-1

1. Answers vary.   3. Answers vary.

5. 4 pints   7. 2 pints   9. 3 quarts

11. 546   13. 4,430

Home Link 7-2

1. $\frac{7}{4}$, or $1\frac{3}{4}$ cups

3. a. $\frac{3}{6}$, or $\frac{1}{2}$ cup   b. $\frac{15}{6}$, or $2\frac{3}{6}$, or $2\frac{1}{2}$ cups

5. 3,250   7. 22,104

Home Link 7-3

1. $4 * \frac{1}{5} = \frac{4}{5}; \frac{4}{5}$

3. $5 * \frac{1}{2} = \frac{5}{2}$, or $2\frac{1}{2}; \frac{5}{2}$, or $2\frac{1}{2}$ avocados

5. $\frac{3}{2}$, or $1\frac{1}{2}$ 7. $\frac{5}{10}$, or $\frac{1}{2}$

Home Link 7-4

1. $\frac{5}{5}$, or 1   3. $\frac{18}{6}$, or 3

5. $5 * \frac{6}{10} = \frac{30}{10}$, or 3 miles

   $7 * \frac{6}{10} = \frac{42}{10}$, or $4\frac{2}{10}$, or $4\frac{1}{5}$ miles

7. 2,096   9. 14,752

Home Link 7-5

1. $5 * 1\frac{1}{2} = l$; $\frac{15}{2}$, or $7\frac{1}{2}$ pounds; 7 and 8; 120 ounces

3. $14\frac{3}{6}$; 14 and 15

5. $\frac{6}{4}$, or $1\frac{2}{4}$ 7. $\frac{3}{6}$

Home Link 7-6

1. $8 * \frac{3}{8} = \frac{24}{8}$, or 3 pounds

3. $4 * \frac{5}{8}$ lb $= \frac{20}{8}$, or $2\frac{4}{8}$ pounds

5. 45 R1   7. 192 R3

Home Link 7-7

1. A; \$2 more per ticket; Sample answer: $276 \div 2 = 138$; $138 \div 6 = 23$; $336 \div 2 = 168$; $168 \div 8 = 21$

3. 4,524   5. 5,817

Home Link 7-8

1. Sample answer: $(5 * 1,000) - (8 * 500) = w$; 1,000 milliliters

3. Sample answer: $1,400 - (13 * 100) = p$; 100 centimeters

5. $3\frac{4}{6}$ 7. $5\frac{2}{12}$

Home Link 7-9

1. The perimeter is 4 times the side length.

3. 125 toothpicks

5. 251   7. 31 R4

Home Link 7-10

1. **a.** Yes; $\frac{5}{2}$, or $2\frac{1}{2}$ miles

   b. $\frac{10}{2}$, or 5 miles; Sample answer: Tony will run $\frac{1}{2}$ mile 5 times a week. $5 * \frac{1}{2} = \frac{5}{2}$ miles. For 2 weeks, add $\frac{5}{2} + \frac{5}{2} = \frac{10}{2}$, or 5 miles. $5 > 4$.

3. 321   5. 147 R4

Home Link 7-11

1. $\frac{15}{5}$, or 3   3. 9

5. $1\frac{2}{4}$ pounds; 24 ounces

7. 116 R2   9. 42 R1

Home Link 7-12

1. \$5.53; Sample answer: $7 * \frac{79}{100} = \frac{79}{100} + \frac{79}{100} + \frac{79}{100} + \frac{79}{100} + \frac{79}{100} + \frac{79}{100} + \frac{79}{100} = \frac{553}{100} = 5$ and 53 hundredths $= 5.53$

3. \$1.69; Sample answer: $\frac{1,000}{100} - \frac{831}{100} = \frac{169}{100}$

5. $=$ 7. $>$

Home Link 7-13

Pencil Lengths (to the nearest $\frac{1}{8}$ inch)

1. 13 students

3. **a.** 3 pencils   **b.** 6 inches

5. **a.** $3\frac{4}{8}$ inches   **b.** $1\frac{7}{8}$ inches

   c. $4\frac{11}{8}$, or $5\frac{3}{8}$ inches   **d.** $1\frac{5}{8}$ inches

7. $12\frac{2}{10}$ 9. $3\frac{90}{100}$

Liquid Measures

Find at least one container that holds each of the amounts listed below.
Describe each container and record all the measurements on the label.

SRB
195

① About 1 gallon

Container	Liquid Measurements on Label
jug of orange juice	*gallon, 3.78 L*

② About 1 quart

Container	Liquid Measurements on Label
container of milk	*1 quart, 32 fl oz*

③ About 1 pint

Container	Liquid Measurements on Label

④ About 1 cup

Container	Liquid Measurements on Label

Complete.

⑤ 2 quarts = _____ pints

⑥ 3 gallons = _____ cups

⑦ _____ pints = 4 cups

⑧ _____ quarts = 12 cups

⑨ 6 pints = _____ quarts

⑩ _____ quarts = $2\frac{1}{2}$ gallons

Practice

⑪ 273 * 2 = _____

⑫ 385 * 4 = _____

⑬ _____ = 886 * 5

⑭ _____ = 98 * 38

Sugar in Drinks

Use the information in the table to solve the number stories. In the space below each problem, use pictures or equations to show what you did to find your answers.

SRB
160-161,
196-197

Amount of Sugar in Drinks		
Drink	**Sugar Content (in cups)**	**Serving Size (in ounces)**
Cranberry juice cocktail	$\frac{1}{4}$	12
Fruit punch	$\frac{1}{4}$	12
Orange soda	$\frac{1}{4}$	12
Sweet tea	$\frac{1}{6}$	12

Sources: National Institutes of Health and California Department of Public Health

① Carmen drinks one 12-ounce can of orange soda every day. How much sugar is that in 1 week? _____ cup(s)

② If you drink one 12-ounce glass of cranberry juice cocktail every morning, how much sugar will that be in 2 weeks? _____ cup(s)

③ Mike drinks three 12-ounce servings of sweet tea per day.

    a. How much sugar is he drinking in his tea in 1 day?

    _____ cup(s)

    b. In 5 days? _____ cup(s)

Practice

④ $951 * 4 =$ _____

⑤ $650 * 5 =$ _____

⑥ $425 * 7 =$ _____

⑦ $3,684 * 6 =$ _____

Multiplying Unit Fractions

Write a multiplication equation to describe each picture or story.

(1)

Multiplication equation: _____

What is the fourth multiple of $\frac{1}{5}$? _____

(2)

Multiplication equation: _____

What is the third multiple of $\frac{1}{10}$? _____

(3) Dmitri fixed a snack for 5 friends. Each friend got $\frac{1}{2}$ of an avocado. How many avocados did Dmitri use?

Multiplication equation: _____

Answer: _____ avocado(s)

(4) Juanita made 3 protein shakes. All together, she used 1 cup of protein powder to make them. Each had the same amount.

How many cups of protein powder are in each shake?

Multiplication equation: _____

Answer: _____ cup(s)

Practice

(5) $\frac{1}{2} + \frac{1}{2} + \frac{1}{2} =$ _____

(6) $\frac{2}{3} + \frac{2}{3} + \frac{1}{3} =$ _____

(7) $\frac{9}{10} - \frac{4}{10} =$ _____

(8) $\frac{8}{12} - \frac{5}{12} =$ _____

Multiplying Fractions by Whole Numbers

Solve the problems below.

(1) $5 * \frac{1}{5} =$ _____

Draw a picture.

(2) $3 * \frac{4}{9} =$ _____

Draw a picture.

(3) $6 * \frac{3}{6} =$ _____

Draw a picture.

Write a multiplication equation to represent the problem and then solve.

(4) Rahsaan needs to make 5 batches of granola bars. A batch calls for $\frac{1}{2}$ cup of honey.

How much honey does he need? Equation: _____

(5) Joe swims $\frac{6}{10}$ of a mile 5 days per week. How far does he swim every week?

Equation: _____

How far would he swim if he swam every day of the week?

Equation: _____

Practice

(6) $653 * 3 =$ _____

(7) $262 * 8 =$ _____

(8) $357 * 9 =$ _____

(9) $7,376 * 2 =$ _____

Multiplying Mixed Numbers by Whole Numbers

Solve.

(1) Michelle's grandmother sent her 5 small gifts for her fifth birthday. Each one weighed $1\frac{1}{2}$ pounds. How much did the gifts weigh all together?

Number model with unknown: _____

Answer: _____ pounds

Between what two whole numbers is this? _____ and _____

How many ounces did the gifts weigh? _____ ounces

(2) Rochelle bought 4 pieces of ribbon to finish a project. Each piece was $1\frac{5}{12}$ yards long. What is the combined length of the ribbon she bought?

Number model with unknown: _____

Answer: _____ yards

Between what two whole numbers is this? _____ and _____

How many feet is this? _____ feet

(3) $3 * 4\frac{5}{6} = $ _____

Between what two whole numbers is this? _____ and _____

(4) $6 * 7\frac{3}{8} = $ _____

Between what two whole numbers is this? _____ and _____

Practice

(5) $\frac{3}{4} + \frac{2}{4} + \frac{1}{4} = $ _____

(6) $\frac{4}{8} + \frac{3}{8} + \frac{2}{8} = $ _____

(7) $\frac{5}{6} - \frac{2}{6} = $ _____

(8) $\frac{88}{100} - \frac{57}{100} = $ _____

Fruit Salad Weight

Mr. Chou makes fruit salad that he sells in his store. Today he plans to make a fruit salad with 8 pears, 2 cups of grapes, and 4 pints of strawberries. Use the weights below to solve the problems.

- A medium pear weighs about $\frac{3}{8}$ lb.

- A cup of grapes weighs about $\frac{2}{8}$ lb.

- A pint of strawberries weighs about $\frac{5}{8}$ lb.

① Write a multiplication sentence to show how much the pears weigh. _____

Answer: _____ pound(s)

② Write a multiplication sentence to show how much the grapes weigh. _____

Answer: _____ pound(s)

③ Write a multiplication sentence to show how much the strawberries weigh.

_____

Answer: _____ pound(s)

④ How much does Mr. Chou's salad weigh in all? Show your work.

Answer: _____ pound(s)

Practice

⑤ 361 / 8 = _____

⑥ 396 ÷ 7 = _____

⑦ 963 / 5 = _____

⑧ 633 / 4 = _____

Division Number Stories

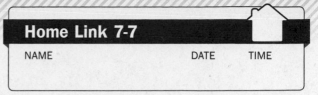
Solve. Show your work.

SRB
111-115

(1) Robert and Jason want to buy a group ticket package for football games. Package A costs $276 and includes 2 tickets for each of 6 games. Package B costs $336 and includes 2 tickets for each of 8 games. Which package charges more per ticket? How much more per ticket?

Package _____ charges $_____ more per ticket.

(2) Rebecca wants to put 544 pennies in a coin-collection book. The blue book fits 9 pennies per page. The red book fits 7 pennies per page. How many more pages would she need if she used the red book rather than the blue one?

The red book will take _____ more pages than the blue book.

What did you do with any remainders you found?

_____

_____

_____

Practice

(3) 754 * 6 = _____

(4) 906 * 2 = _____

(5) _____ = 831 * 7

(6) _____ = 84 * 29

203

More Division Measurement Number Stories

Read each number story. Use the information to write a number model with an unknown and then solves.

① Kelly is in charge of bringing water for her softball game. The 8 members of the team have matching team water bottles that hold 500 mL. Kelly buys 5 liters of water at the store. If she fills all the bottles, how many milliliters of water will Kelly have left?

Number model with unknown: _____

Answer: _____ milliliters

② The distance around all the bases in softball is 72 meters. If Kelly hits 2 home runs and runs around the bases twice, how many millimeters will she run?

Number model with unknown: _____

Answer: _____ millimeters

③ In women's softball the pitcher stands about 13 meters from the batter's box. In men's softball the pitcher stands about 1,400 centimeters from the batter's box. About how many more centimeters is it from the men's pitcher to the batter's box than from the women's pitcher to the batter's box?

Number model with unknown: _____

Answer: About _____ centimeters

④ The 6 games Kelly's team played took a total of 7 hours.

a. How many minutes total did they play softball?

Number model with unknown: _____

Answer: _____ minutes

b. If each game lasted the same amount of time, how many minutes did each one last?

Number model with unknown: _____

Answer: _____ minutes

Practice

⑤ $1\frac{3}{6} + 2\frac{1}{6} =$ _____

⑥ $4\frac{3}{5} + 5\frac{4}{5} =$ _____

⑦ $7\frac{5}{12} - 2\frac{3}{12} =$ _____

⑧ $6\frac{1}{3} - 2\frac{2}{3} =$ _____

Perimeter Patterns

Alice was making squares out of toothpicks. She noticed a pattern involving the length of one side and the perimeter of the square. Complete the table and then answer the questions that follow.

SRB
200-201,
58-59

Side Length	Perimeter
1	4
2	
	12
4	
	20

① What rule describes the relationship between the length of one side and the perimeter of a square?

_____

② What would be the perimeter of a square with a side length of 25 toothpicks?

_____ toothpicks

③ What would be the side length of a square with a perimeter of 500 toothpicks?

_____ toothpicks

④ Describe at least two other patterns you notice in the table _____

_____

_____

_____

_____

_____

_____

Practice

⑤ $753 \div 3 =$ _____

⑥ _____ $= 386 \div 2$

⑦ $283 \div 9 \rightarrow$ _____

⑧ $505 \div 6 \rightarrow$ _____

Fitness Challenge

Use the information in the table below to solve the number stories.

During Marcy School's 2-week challenge, each student who meets a goal wins a prize.

SRB
162-164,
173-174

Marcy's Fitness Challenge Goals			
Activity	**Total Distance**	**Activity**	**Total Distance**
Walking	6 miles	Bike Riding	6 miles
Swimming	1 mile	Running	4 miles

(1) Tony will run $\frac{1}{2}$ mile after school each day. Will he win a prize? _____

a. Distance run in 1 week: _____ mile(s)   b. In 2 weeks: _____ mile(s)

Explain how you found your answer.

_____

_____

(2) Three times a week, Tina walks $\frac{3}{10}$ mile from school to the library, studies for 1 hour, and then walks $\frac{4}{10}$ mile home. How much more will she need to walk to win a prize?

_____ mile(s)

Explain how you found your answer.

_____

_____

_____

Practice

(3) $642 \div 2 =$ _____

(4) $386 / 9 \rightarrow$ _____

(5) $739 / 5 \rightarrow$ _____

(6) $4\overline{)829} \rightarrow$ _____

Fractions and Mixed Numbers

Solve. Draw a picture or show how you solved the problem.

(1) $5 * \frac{3}{5} =$ _____

(2) _____ $= 4\frac{2}{6} - 2\frac{4}{6}$

(3) $5\frac{7}{8} + 3\frac{1}{8} =$ _____

(4) _____ $= 3 * 4\frac{1}{4}$

(5) The combined weight of an assortment of fruit is $8\frac{3}{4}$ pounds. When the fruit is on a tray, the tray weighs $10\frac{1}{4}$ pounds. How many pounds does the tray weigh when empty? _____ pound(s)

How many ounces does the tray weigh when empty? _____ ounce(s)

(6) $\left(3 * 2\frac{2}{3}\right) + \left(2 * 4\frac{1}{3}\right) =$ _____

Practice

(7) $3\overline{)350}$

(8) $6\overline{)832}$

(9) $7\overline{)295}$

(10) $9\overline{)582}$

Shopping for Bargains

Solve each number story and show how you solved the problems.

SRB
166-167,
175-176

① Phil wants to buy some Creepy Creature erasers that cost $1.05 each.
If he buys 5 or more, the price is $0.79 each. If he decides to buy 7 erasers,
how much will he spend?

  Answer: $_____

② Mrs. Katz bought 3 pounds of apples and a muffin for snacks. The apples cost $2.59
per pound if you buy less than 3 pounds and $2.12 per pound if you buy 3 or more
pounds. The muffin cost $1.95. How much did she spend?

  Answer: $_____

Try This

③ Mrs. Katz paid with a $10 bill. How much change did she get back?

  Answer: $_____

Practice

Fill in the blanks with >, <, or =.

④ 0.55 ____ 0.65   ⑤ 0.3 ____ 0.30   ⑥ 0.72 ____ 0.8   ⑦ 0.4 ____ 0.31

Pencil Lengths

At the beginning of the year Mrs. Kerry gave each student in her class a new pencil with "Welcome to 4th Grade" written on it. A month later the class measured their pencils to the nearest $\frac{1}{8}$ inch.

Pencil Lengths to the Nearest $\frac{1}{8}$ inch

$2\frac{1}{8}$	$3\frac{1}{8}$	$2\frac{7}{8}$	$2\frac{4}{8}$	$3\frac{3}{8}$	$2\frac{7}{8}$	3	$2\frac{5}{8}$	$2\frac{5}{8}$	$2\frac{7}{8}$	$3\frac{3}{8}$	$2\frac{6}{8}$	$2\frac{4}{8}$
$2\frac{3}{8}$	$2\frac{7}{8}$	$1\frac{7}{8}$	$3\frac{2}{8}$	$2\frac{7}{8}$	$3\frac{4}{8}$	$2\frac{6}{8}$	$2\frac{3}{8}$	$3\frac{1}{8}$	2	$2\frac{4}{8}$	$2\frac{5}{8}$	$3\frac{2}{8}$

Plot the data set on the line plot.

Title: _____

215

Pencil Lengths
(continued)

Use the completed line plot to answer these questions.

(1) How many students have a pencil that is shorter than $2\frac{7}{8}$ inches?

_____ students

(2) What is the most common pencil length? _____ inches

(3) **a.** How many pencils are less than $2\frac{2}{8}$ inches long? _____ pencils

   b. What is their combined length? _____ inches

(4) **a.** How many pencils are between $2\frac{7}{8}$ and $3\frac{2}{8}$ inches long? _____ pencils

   b. What is their combined length? _____ inches

(5) **a.** How long is the longest pencil? _____ inches

   b. How long is the shortest pencil? _____ inches

   c. What is the combined length of the longest and shortest pencils? _____ inches

   d. What is the difference in length of the longest and shortest pencils?

   _____ inches

Practice

(6) $2\frac{1}{4} + 5\frac{2}{4} =$ _____

(7) $8\frac{5}{10} + 3\frac{7}{10} =$ _____

(8) $3\frac{7}{8} - 1\frac{3}{8} =$ _____

(9) $7\frac{41}{100} - 3\frac{51}{100} =$ _____

Fraction Operations; Applications

Think back to how you learned to ride a bike as a child. What if you were allowed to practice only on a stationary bike rather than a real one? When you finally ventured out onto the neighborhood streets expecting to ride like a pro, you would probably be disappointed! Without an opportunity to apply what you learned to a real-world situation, you would never have to apply the brakes going down a hill or maneuver around a sharp curve. Likewise, if students aren't given a chance to apply what they learn in mathematics to real-world situations, it may seem to them like useless knowledge. To help make mathematics more meaningful to students, Unit 8 asks them to apply what they have learned throughout the year to real-world problems.

Fraction Operations

This year students have explored adding, subtracting, and multiplying fractions. In Unit 8 they will apply fraction and mixed-number operations to help them solve real-world problems involving the perimeter and area of rectangles and units of measure. For example, students will use the relationship between perimeter and area to find the missing side length of a fence or determine the fractional amounts of juice needed to make fruit punch.

Angle Applications

Angles play important roles in many real-life situations, including carpentry, measuring the angles of the sun, and many sports. Lesson 8-2 uses hockey to demonstrate real-world applications of students' knowledge of angles. For instance, when a hockey player wants to pass the puck and an opponent is blocking the path, the passer hits the puck off the boards at an angle, causing the puck to travel around the opponent. This is called "banking the puck." In Lesson 8-2 students also use what they have learned about angles to explore the role angles play in our field of vision, which is the angle that includes the area that can be seen without moving the head or eyes.

More Applications

In Lesson 8-4 students apply their knowledge of symmetry to quilting patterns and then create their own quilt based on specified numbers of lines of symmetry. In Lesson 8-5 students use real-world data about envelope sizes from the U.S. Postal Service to create line plots. They then answer questions about the data by adding and subtracting fractions. In Lesson 8-12 students use their knowledge of place value, addition, and subtraction to solve challenging puzzles called cryptarithms. In Lesson 8-13 students find equivalent names for numbers.

Please keep this Family Letter for reference as your child works through Unit 8.

Vocabulary

Important terms in Unit 8:

equivalent names Different ways of naming the same number. For example, $2 + 6$, $4 + 4$, $12 - 4$, $18 - 10$, $100 - 92$, $5 + 1 + 2$, eight, VIII, and ~~HHH~~ /// are all equivalent names for 8.

fluid ounce (fl oz) A U.S. customary unit of volume or capacity equal to $\frac{1}{16}$ of a pint, or about 29.6 milliliters.

Do-Anytime Activities

To work with your child on concepts taught in this unit, try these activities:

1. Have your child complete number puzzles found in newspapers, magazines, or online. Discuss with your child how he or she found the solutions.

2. Ask your child to measure a rectangular object such as an envelope, notebook, or room in your home. Have him or her find both the perimeter and the area of the object and then compose a word problem about the measurements.

3. Ask your child to point out items that he or she believes are symmetrical. How many lines of symmetry are there in those items?

4. Have your child point out angles in your home and estimate their measures. Ask your child to add angles together or find missing angles based on these estimates.

5. Show your child a food or beverage container and have him or her locate the liquid volume and convert it to a smaller unit. For instance, a juice box might hold 1 cup of juice, which means it holds 8 fluid ounces of juice.

Building Skills through Games

In Unit 8 students play the following game to increase their understanding of numbers and the properties of numbers. For detailed instructions, see the *Student Reference Book*.

Name That Number See *Student Reference Book*, page 268. This game provides practice representing numbers in different ways, using any or all of the four operations: addition, subtraction, multiplication, and division.

As You Help Your Child with Homework

As your child brings assignments home, you may want to go over instructions together, clarifying them as necessary. The answers listed below will guide you through the Home Links in Unit 8.

Home Link 8-1

1. Team B's car; 27 cm   **3.** 180 cm

5. 2,833 R1   **7.** 715 R3

Home Link 8-2

1. 165°; 82° + 83° = f **3.** 87°; 3° + w = 90°

5. 137°; 180° − 43° = s **7.** $\frac{5}{3}$, or $1\frac{2}{3}$

9. $\frac{11}{5}$, or $2\frac{1}{5}$

Home Link 8-3

1. 60°; Sample answers: 30° + 30° = 60°; The measure of each small white rhombus angle is 30°, so two of them make 60°.

3. 16,764   **5.** 4,888

Home Link 8-4

1.

3.

5. 1 line of symmetry;

7. $\frac{30}{6}$, or 5   **9.** $\frac{28}{10}$, or $2\frac{8}{10}$

Home Link 8-5

Book Heights

Height (inches)

1. $2\frac{7}{8}$ in.   **3.** 1,172 R3

Home Link 8-6

1. $7\frac{2}{6}$ yd   **3.** Width = $\frac{12}{100}$ km

5. Width = $2\frac{3}{10}$ cm   **7.** $\frac{4}{3}$, or $1\frac{1}{3}$

9. $\frac{36}{5}$, or $7\frac{1}{5}$

Home Link 8-7

1. 3.26 kilograms; Sample answer: I thought about what number added to 4 would give me $7\frac{26}{100}$. First I added 3 to get 7. Then I added $\frac{26}{100}$ to get $7\frac{26}{100}$. Finally, $3 + \frac{26}{100} = 3\frac{26}{100} = 3.26$

3. 7.8 cm; Sample answer: $11.4 = 11\frac{4}{10}$ and $3.6 = 3\frac{6}{10}$; $11\frac{4}{10} = 10 + \frac{10}{10} + \frac{4}{10} = 10\frac{14}{10}$; $10\frac{14}{10} - 3\frac{6}{10} = 7\frac{8}{10} = 7.8$

5. 14,316   7. 2,016

Home Link 8-8

1. a. $4\frac{1}{12}$ square feet b. $12\frac{8}{12}$ feet

2. a. $5\frac{6}{12}$ square feet b. $6\frac{4}{12}$ feet

3. $8\frac{4}{10}$ square inches

5. $\frac{4}{6}$ 7. $\frac{4}{10}$

Home Link 8-9

1. $5\frac{1}{4}$ feet; Sample answer: $3 * 1\frac{3}{4} = (3 * 1) + (3 * \frac{3}{4}) = 3 + \frac{9}{4} = 3\frac{9}{4}$, or $5\frac{1}{4}$

2. Yes. Sample answer: $(5 * 1\frac{1}{2}) + (4 * 1\frac{3}{4}) = 5\frac{5}{2} + 4\frac{12}{4} = 7\frac{1}{2} + 7 = 14\frac{1}{2}$

3. $\frac{6}{6}$, or 1   5. $\frac{54}{100}$

Home Link 8-10

1. Rule: $* 8$

in (gallons)	out (pints)
2	16
$3\frac{1}{2}$	28
6	48
$7\frac{1}{4}$	58
10	80

3. a. Yes. Sample answer: The total amount of all the ingredients combined is 18 fluid ounces, so the smoothie will fit in the 24-fluid ounce glass.

b. $\frac{3}{4}$ cup

c. $2\frac{1}{4}$ cups orange juice; 12 fluid ounces cold water; 3 cups vanilla ice cream

d. 54 fluid ounces

5. 1,859   7. 519

Home Link 8-11

1. a. $3\frac{1}{8}$ pounds; Sample answer: $\left(1\frac{1}{2} + \frac{1}{2}\right) + \left(\frac{3}{4} + \frac{1}{4}\right) + \frac{1}{8} = 2 + 1 + \frac{1}{8} = 3\frac{1}{8}$

b. 50 ounces; Sample answer: One pound equals 16 ounces; $\frac{1}{8}$ of a pound $= 2$ ounces; so $(3 * 16) + 2 = 48 + 2 = 50$

c. 2 packages; Sample answer: 1 of each size uses 50 ounces, so 2 of each size would use $2 * 50 = 100$ ounces. $100 > 80$, so 1 package isn't enough.

2. $1\frac{2}{8}$, or $1\frac{1}{4}$ pounds; Sample answer: $\left(\frac{1}{8} + \frac{1}{8}\right) + \left(\frac{1}{4} + \frac{3}{4}\right) = \frac{2}{8} + 1 = 1\frac{2}{8}$, or $1\frac{1}{4}$

3. 15,321   5. 2,146

Home Link 8-12

1. Sample answer: $973 + 51 = 1,024$

3. $80 * 64 = 5,120$

5. a. 27; $9 * 3 = 27$ b. $\frac{1}{3}$; $3 / 9 = \frac{1}{3}$

7. $4\frac{10}{8}$, or $5\frac{2}{8}$ 9. $10\frac{181}{100}$, or $11\frac{81}{100}$

Home Link 8-13

1. Sample answers:

9,990
$2,016 + 7,974$
$(1,427 * 7) + 1$
$1,665 * 6$
$9,000 + 900 + 90$
$13,558 - 3,568$

3. Answers vary.

5. $3\frac{2}{4}$ 7. $2\frac{8}{12}$

Multistep Number Stories

SRB
26, 111-
114

The fourth-grade students in Mr. Kennedy's class are investigating energy and motion. Students worked in teams to build two machines: a car that is propelled by a mousetrap and a boat that is propelled by balloons. Today the teams are competing to see which cars and boats go farthest.

Each car or boat gets 3 trials. The total distance from all 3 trials is used to determine which car or boat went farthest. Solve the number stories to help Mr. Kennedy's class compare the machines made by various teams.

① Team A's car went 173 cm on the first trial, 206 cm on the second trial, and 245 cm on the third trial. Team B's car went 217 cm on each of the three trials.

Which car went the farthest overall? _____

How much farther did it go? _____

② Team A's boat went 130 cm in all. Team B's boat went the same distance on all 3 trials and lost to Team A's boat by 7 cm.

How far did Team B's boat go on each trial? _____

③ Team D's car went the same distance on each of its trials. Team C's car went exactly 1 cm farther in each trial than Team D's car. Team C's car went 543 cm in all.

How far did Team D's car go on each trial? _____

Practice

④ $5,624 \div 8 =$ _____

⑤ $8,500 \div 3 =$ _____

⑥ $4\overline{)9,207}$

⑦ $5\overline{)3,578}$

221

Finding Unknown
Angle Measures

Find the missing angle measures. For each problem, write an equation with a letter for the unknown to show how you found your answer.

① 82° 83°

f

$f =$ _____

Equation: _____

② p

210°

$p =$ _____

Equation: _____

③ 3°

w

$w =$ _____

Equation: _____

④ 75° l

$l =$ _____

Equation: _____

⑤ 43° s

$s =$ _____

Equation: _____

⑥ 120°

W Addison St.

i

N Clark St.

$i =$ _____

Equation: _____

Practice

⑦ $\frac{1}{3} + \frac{2}{3} + \frac{2}{3} =$ _____

⑧ $\frac{1}{4} + \frac{3}{4} + \frac{3}{4}$ _____

⑨ $\frac{4}{5} + \frac{4}{5} + \frac{3}{5} =$ _____

⑩ $\frac{5}{12} + \frac{3}{12} + \frac{7}{12} =$ _____

Finding Pattern-Block Measures

Molly is using pattern blocks to find angle measures of other pattern blocks. She knows that the measure of the small angle of a white rhombus is 30°.

30°   30°

SRB
211-212

(1) Molly fills an angle of the green triangle with the small angles of white rhombuses. What is the measure of the triangle's angle? Explain how you know.

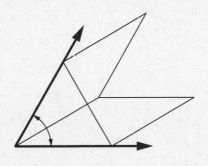

Angle measure: _____°

(2) Molly fills a red trapezoid's large angle with angles of the green triangle. What is the measure of the red trapezoid's large angle? Explain how you know.

Angle measure: _____°

Practice

(3) 5,588 * 3 = _____

(4) 9,037 * 5 = _____

(5) 52 * 94 = _____

(6) 83 * 77 = _____

Line Symmetry

Use a straightedge to draw the lines of symmetry on each shape.

SRB
238

(1) Draw 2 lines of symmetry.

(2) Draw 6 lines of symmetry.

(3) Draw 1 line of symmetry.

(4) Draw 3 lines of symmetry.

(5) How many lines of symmetry does this shape have? _____

Draw the line(s) of symmetry.

(6) Draw your own shape. Show the lines of symmetry. Be sure your shape includes at least 1 right angle.

Practice

(7) $6 * \frac{5}{6} =$ _____

(8) $3 * \frac{3}{8} =$ _____

(9) $4 * \frac{7}{10} =$ _____

(10) $6 * \frac{4}{12} =$ _____

227

Designing a Bookcase

Nicholas is building a bookcase. To help with the design, he measured the height of each of his books to the nearest $\frac{1}{8}$ inch. His measurements are given below.

SRB
215-216

$$6\frac{1}{2}, 9\frac{1}{4}, 7\frac{1}{8}, 7\frac{1}{2}, 8, 6\frac{7}{8}, 9\frac{1}{4}, 9\frac{1}{4}, 9\frac{1}{4}, 9\frac{1}{4}, 9\frac{1}{4}, 8\frac{1}{4}, 8, 8\frac{1}{4}, 8\frac{3}{8},$$

$$6\frac{1}{2}, 7\frac{1}{8}, 9, 6\frac{7}{8}, 9\frac{3}{8}, 6\frac{7}{8}, 7\frac{1}{2}, 8, 8\frac{1}{4}, 9\frac{1}{4}, 6\frac{7}{8}, 6\frac{7}{8}, 8\frac{1}{4}, 8\frac{1}{4}, 8\frac{1}{4}$$

Plot the data set on the line plot below.

Book Heights

Height (inches)

Use the completed line plot to answer the questions below.

1. What is the difference in height between the tallest and shortest books? _____ in.

2. Nicholas wants the space between the shelves to be $\frac{7}{8}$ inch taller than his tallest book.

   a. How far apart should he make the shelves? _____ in.

   b. If the thickness of the wood he uses for the shelves is $\frac{5}{8}$ inch, what will be the total height of each shelf? (*Hint:* The total height is the thickness of one piece of wood plus the distance between shelves.) _____ in.

Practice

3. $8,207 \div 7 \rightarrow$ _____

4. $7,109 \div 8 \rightarrow$ _____

Perimeters and Missing Measures

Use a formula to find the perimeter of each rectangle. Show your work in the space provided.

① Length = $3\frac{3}{6}$ yd
Width = $\frac{1}{6}$ yd

Perimeter: _____ yd

② Length = $5\frac{1}{12}$ ft

Width = $4\frac{11}{12}$ ft

Perimeter: _____ ft

For each rectangle, find the unknown side measure.

③ Perimeter: $\frac{74}{100}$ kilometer

Length = $\frac{25}{100}$ km

Width = _____ km

④ Perimeter: 10 inches

Length = $4\frac{3}{8}$ in.
Width = _____ in.

⑤ Perimeter: $12\frac{8}{10}$ centimeters

Length = $4\frac{1}{10}$ cm

Width = _____ cm

Try This

⑥ Perimeter: $16\frac{1}{2}$ ft

Length = _____ ft

Width = $3\frac{1}{2}$ ft

Practice

⑦ $2 * \frac{2}{3} =$ _____

⑧ $5 * \frac{3}{4} =$ _____

⑨ $9 * \frac{4}{5} =$ _____

⑩ $8 * \frac{6}{12} =$ _____

Decimal Number Stories

Solve each number story. Write your answer as a decimal.
Show how you found your answer.

SRB
166-167

(1) An Olympic men's shot put weighs 7.26 kilograms. An Olympic women's shot put weighs 4 kilograms. How much more does the men's shot put weigh than the women's shot put?

_____ kilograms

(2) The recipe for homemade glue calls for 0.5 liter of skim milk, 0.09 liter of vinegar, and 0.06 liter of water. When you combine the ingredients, how much liquid will you have?

_____ liter

(3) Ben cut a piece of string 11.4 cm long. Then he cut 3.6 cm off of it. How long is the string now?

_____ cm

Try This

(4) What is the answer to Problem 3 in milliliters? _____ milliliters

Practice

(5) $3{,}579 * 4 =$ _____

(6) $2{,}904 * 6 =$ _____

(7) $36 * 56 =$ _____

(8) $47 * 72 =$ _____

Area and Perimeter

SRB
175-204

Solve the problems below.

1. The Murphy family bought two rectangular dog beds for their pets. Fluffy's bed was 3 feet by $1\frac{9}{12}$ feet. Pete's bed was 4 feet by $2\frac{4}{12}$ feet.

   a. How much more area does Pete's bed have than Fluffy's?

   Answer: _____ square feet

   b. What is the perimeter of Pete's bed? Answer: _____ feet

2. The Cho family bought two rectangular cat beds for their cats. George's bed is 2 feet by $1\frac{2}{12}$ feet. Sammie's bed is 2 feet by $1\frac{7}{12}$ feet.

   a. What is the total area of these two beds? Answer: _____ square feet

   b. What is the perimeter of George's bed? Answer: _____ feet

3. Perimeter: $12\frac{2}{10}$ inches

   x inches

   $2\frac{1}{10}$ inches

   Area: _____ square inches

4. Area: $9\frac{3}{8}$ square feet

   x feet

   3 feet

   Width: _____ feet

Practice

5. $\frac{5}{6} - \frac{1}{6} =$ _____

6. $\frac{8}{8} - \frac{3}{8} =$ _____

7. $\frac{9}{10} - \frac{5}{10} =$ _____

8. $\frac{11}{12} - \frac{5}{12} =$ _____

Using Doghouse Dimensions

Dan and Diane's Doghouse Dynasty builds doghouses to order. They can change the
length and width for doghouses, but they always build them to have the same height.
Solve the number stories about doghouses built to certain widths and lengths based
on the information given in the table. Use drawings or equations to show how you solved
each problem.

SRB 162, 175

Custom Doghouse Dimensions		
Size	Length (in feet)	Width (in feet)
Extra small	$3\frac{1}{4}$	$1\frac{1}{3}$
Small	$3\frac{1}{2}$	$1\frac{1}{2}$
Medium	4	$1\frac{3}{4}$
Large	$4\frac{1}{4}$	$1\frac{5}{6}$
Extra large	$4\frac{5}{6}$	2

width   length

① Mrs. Swift ordered 3 medium-size doghouses. What will their combined width be?

_____ feet

② Kisa's Kennel has a space that is 18 feet wide in which they want to place
doghouses side by side. If they order 5 small and 4 medium doghouses, will
they all fit in the space? _____

Practice

③ $2 * \frac{3}{6} =$ _____

④ $5 * \frac{7}{10} =$ _____

⑤ $9 * \frac{6}{100} =$ _____

⑥ $7 * \frac{8}{12} =$ _____

237

Liquid Measurement and Fractions

Complete the "What's My Rule?" tables and state the rules.

SRB
196-197

1 Rule: _____

in (gallons)	out (pints)
2	16
$3\frac{1}{2}$	
	48
$7\frac{1}{4}$	
	80

2 Rule: _____

in (quarts)	out (cups)
3	12
$4\frac{1}{2}$	
	32
$9\frac{3}{4}$	
$12\frac{1}{4}$	

Use this recipe for a Creamsicle Smoothie to solve the problems below.

$\frac{3}{4}$ cup orange juice   4 fluid ounces cold water   1 cup vanilla ice cream

Combine all ingredients.

3   **a.** Will this recipe fit in a glass that holds 24 fluid ounces? _____

      Explain your thinking. _____

      _____

  b. About how many more cup(s) of smoothie could fit in the glass? _____ cup(s)

  c. Frank wants to triple the recipe. How much of each ingredient will he need?

      _____ orange juice

      _____ cold water

      _____ vanilla ice cream

  d. After tripling the recipe, how much smoothie will Frank have? _____ fluid ounces

Practice

4 3,560 ÷ 3 → _____

5 9,295 ÷ 5 → _____

6 7)8,210

7 9)4,671

239

Planning a Cookout

The Whispering Lakes Neighborhood Association is having a hamburger cookout. Each family can choose whether to order the hamburgers or bring their own. Use the information in the table to solve the number stories. Use drawings, tables, or equations to show what you did.

Size of Hamburger	Weight of One Hamburger Patty (lb)
Small	$\frac{1}{8}$
Medium	$\frac{1}{4}$
Large	$\frac{1}{2}$
Jumbo	$\frac{3}{4}$
King of the Burgers	$1\frac{1}{2}$

SRB
190-191

(1) **a.** What is the combined weight of 1 of each size hamburger?

_____ pounds

b. How many ounces is that?

_____ ounces

c. Mrs. Ward found 80-ounce packages of hamburger on sale. If she needs to make 2 of each size hamburger, how many packages of meat will she need to buy?

_____ packages

(2) The Finch family ordered 2 small hamburgers, 1 medium hamburger, and 1 jumbo hamburger. How many pounds of hamburger meat does the neighborhood association need to buy for this family?

_____ pounds

Practice

(3) 5,107 * 3 = _____

(4) 4,794 * 6 = _____

(5) 74 * 29 = _____

(6) 93 * 48 = _____

241

Number-Tile Computations

Cut out the 0–9 number tiles at the bottom of the page. Use them to help you solve the problems. Each of the 20 tiles can only be used once.

SRB
92-95

① Use odd-numbered tiles 1, 3, 5, 7, and 9 to make the largest sum.

② Use even-numbered tiles 0, 2, 4, 6, and 8 to make the smallest difference.

③ Use number tiles 0, 4, 6, and 8 to make the largest product.

④ Use number tiles 1, 2, 5, and 7 to make the smallest whole-number quotient. The answer may have a remainder.

⑤ Answer the following questions using only the unused tiles and any operation. Write number sentences to show your work.

   a. What is the largest answer you can find? _____

   ☐ ___ ☐ = ___

   b. What is the smallest answer you can find? _____

   ☐ ___ ☐ = ___

Practice

⑥ $4\frac{3}{5} + 3\frac{4}{5} =$ _____

⑦ $1\frac{5}{8} + 3\frac{5}{8} =$ _____

⑧ $2\frac{9}{12} + 4\frac{5}{12} =$ _____

⑨ $5\frac{89}{100} + 5\frac{92}{100} =$ _____

0	1	2	3	4	5	6	7	8	9
0	1	2	3	4	5	6	7	8	9

Many Names for Numbers

Write five names in each box below. Use as many different kinds of numbers (such as whole numbers, fractions, decimals) and different operations (+, −, *, ÷) as you can.

SRB
33

①
9,990

②
32.68

Make up your own name-collection boxes.

③

④

Practice

⑤ $5\frac{1}{4} - 1\frac{3}{4} =$ _____

⑥ $4\frac{3}{10} - 2\frac{7}{10} =$ _____

⑦ $6\frac{7}{12} - 3\frac{11}{12} =$ _____

⑧ $8\frac{1}{6} - 4\frac{5}{6} =$ _____

245

Congratulations!

By completing *Fourth Grade Everyday Mathematics*, your child has accomplished a great deal. Thank you for all of your support this year.

This Family Letter is a resource to use throughout your child's vacation. It includes an extended list of "Do-Anytime Activities," directions for games that can be played at home, a list of mathematics-related books to check out over vacation, and a sneak preview of what your child will be learning in *Fifth Grade Everyday Mathematics*. Enjoy your vacation!

Do-Anytime Activities

Mathematics means more to everyone when it is rooted in real-life situations. To help your child review many of the concepts he or she has learned in fourth grade, we suggest the following activities for you to do together over the break. These activities will not only help to prevent your child from forgetting content, but they will also help prepare him or her for *Fifth Grade Everyday Mathematics*.

1. Practice multiplication and division facts to maintain fluency.

2. Convert measurements in real-world contexts. For example, at the grocery store ask, "How many quarts are in this gallon of milk?"

3. Have your child practice multidigit multiplication and division using the algorithms with which he or she is most comfortable.

4. Look at advertisements and compare sale prices to original prices. Use a calculator to find unit prices to determine possible savings.

Building Skills through Games

The following section lists rules for games that can be played at home. You will need a deck of number cards, which can be made from index cards or by modifying a regular deck of cards as follows:

A regular deck of playing cards includes 54 cards (52 regular cards plus 2 jokers). Use a permanent marker to write on the cards or a ballpoint pen to write on pieces of white adhesive labels to mark some of the cards:

- Mark each of the four aces with the number "1."

- Mark each of the four queens with the number "0."

- Mark each of the four jacks and the four kings with one of the numbers from 11–18.

- Mark the two jokers with the numbers 19 and 20.

Name That Number

Materials   1 set of cards. See above for directions to make this set.

Players   2 or 3

Object of the Game   To collect the most cards

Directions

1. Shuffle the cards and deal five cards to each player. Place the remaining cards number-side down. Turn over the top card and place it beside the deck. This is the **target number** for the round.

2. Players try to match the target number by adding, subtracting, multiplying, or dividing the numbers on as many of their cards as possible. A card may be used only once.

3. Players write their solutions on a sheet of paper or a slate. When players have written their best solutions, they:

    • Set aside the cards they used to name the target number.

    • Replace used cards by drawing new cards from the top of the deck.

    • Put the old target number on the bottom of the deck.

    • Turn over a new target number and play another hand.

4. Play continues until there are not enough cards left to replace all of the players' cards. The player who sets aside more cards wins the game.

Example: Target number: 16   A player's cards:

Some possible solutions:

$10 + 8 - 2 = 16$ (*three cards used*)

$7 * 2 + 10 - 8 = 16$ (*four cards used*)

$8 / 2 + 10 + 7 - 5 = 16$ (*all five cards used*)

The player sets aside the cards used to make a solution and draws the same number of cards from the top of the deck.

Top-It Games

Materials   Number cards 1–9 (4 of each) as described above
                1 calculator (optional)

Players   2 to 4

Skills   Addition, Subtraction, and Multiplication

Object of the Game   To collect the most cards

Addition Top-It
Directions

1. Shuffle the cards and place them number-side down on the table.

2. Each player takes eight cards, forms two 4-digit numbers, and finds the sum. Players should carefully consider how they form their numbers, because different arrangements lead to different sums. For example, 7,431 + 5,269 has a greater sum than 1,347 + 2,695. The player with the largest sum takes all the cards. In case of a tie, each player turns over eight more cards and calls out the sum. The player with the largest sum takes all the cards from both rounds.

3. Check answers, using a calculator if necessary.

4. The game ends when there are not enough cards left for each player to have another turn.

5. The player with the most cards wins.

Subtraction Top-It
Directions

1. Shuffle the cards and place the deck number-side down on the table.

2. Each player takes eight cards, forms two 4-digit numbers, and finds the difference. Players should carefully consider how they form their numbers, because different arrangements lead to greater differences. For example, 7,431 − 5,269 has a smaller difference than 7,431 − 2,695. The player with the largest difference takes all the cards. In case of a tie, each player turns over eight more cards and calls out the difference. The player with the largest difference takes all the cards from both rounds.

3. Check answers, using a calculator if necessary.

4. The game ends when there are not enough cards left for each player to have another turn.

5. The player with the most cards wins.

Multiplication Top-It
Directions

1. Shuffle the cards and place them number-side down on the table.

2. Each player turns over four cards, forms two 2-digit numbers, and finds the product. Players should carefully consider how they form their numbers, because different arrangements lead to different products. For example, 74 * 52 has a greater product than 47 * 25. The player with the largest product takes all the cards. In case of a tie, each player turns over four more cards and calls out the product. The player with the largest product takes all the cards from both rounds.

3. Check answers, using a calculator if necessary.

4. The game ends when there are not enough cards left for each player to have another turn.

5. The player with the most cards wins.

Vacation Reading with a Mathematical Twist

Books can contribute to students' learning by representing mathematics in a combination of real-world and imaginary contexts. The titles listed below were recommended by teachers who use *Everyday Mathematics* in their classrooms. They are organized by mathematical topic. Visit your local library and check out these and other mathematics-related books with your child.

Operations and Algebraic Thinking
A Remainder of One by Elinor J. Pinczes
17 Kings and 42 Elephants by Margaret Mahy
Anno's Magic Seeds by Mitsumasa Anno
Pattern by Henry Pluckrose
The Grapes of Math by Greg Tang

Numeration and Operations in Base-Ten
If the World Were a Village by David J. Smith
The Doorbell Rang by Pat Hutchins
*The Man Who Counted: A Collection of
  Mathematical Adventures* by Malba Tahan
The Grizzly Gazette by Stuart J. Murphy

Numeration and Operations: Fractions
Fraction Fun by David A. Adler
Working with Fractions by David Adler
Full House by Dayle Ann Dodds
Funny & Fabulous Fraction Stories by Dan
  Greenberg
*Civil War Recipes: Adding and Subtracting
  Simple Fractions* by Lynn George

*Music Math: Exploring Different Interpretations
  of Fractions* by Kathleen Collins
My Half Day by Doris Fisher and Dani Sneed
The Wishing Club by Donna Jo Napoli

Measurement and Data
How Tall, How Short, How Faraway
  by David A. Adler
Is a Blue Whale the Biggest Thing There Is?
  by Robert E. Wells
Math Curse by Jon Scieszka and Lane Smith
Counting on Frank by Rod Clement
Spaghetti and Meatballs for All! by Marilyn Burns

Geometry
The Greedy Triangle by Marilyn Burns
Grandfather Tang's Story by Ann Tompert
Sweet Clara and the Freedom Quilt by Deborah
  Hopkinson
Whale of a Tale by Barbara Pearl
Zachary Zormer, Shape Transformer by Joanne
  Reisberg

Looking Ahead: *Fifth Grade Everyday Mathematics*

Next year, your child will . . .

- Continue to explore and practice whole-number operations, including the use of exponents, and work with larger numbers.

- Expand skills with decimals and fractions, includeing using all four operations.

- Investigate methods for solving problems using mathematics in everyday situations.
  o Graph points on coordinate planes to solve real-world mathematical problems
  o Work with number lines, times, dates, and rates
  o Collect, organize, describe, and interpret numerical data

- Analyze patterns and relationships.

- Further explore the properties, relationships, and measurement of 2-dimensional objects and begin to work with 3-dimensional objects.

- Understand the concepts of volume.

Again, thank you for all of your support this year. Have fun increasing your own understanding of mathematics while continuing your child's mathematical learning!